ARCHAEOBIOLOGY

ARCHAEOLOGIST'S TOOLKIT

SERIES EDITORS: LARRY J. ZIMMERMAN AND WILLIAM GREEN

The Archaeologist's Toolkit is an integrated set of seven volumes designed to teach novice archaeologists and students the basics of doing archaeological fieldwork, analysis and presentation. Students are led through the process of designing a study, doing survey work, excavating, properly working with artifacts and biological remains, curating their materials, and presenting findings to various audiences. The volumes—written by experienced field archaeologists—are full of practical advice, tips, case studies, and illustrations to help the reader. All of this is done with careful attention to promoting a conservation ethic and an understanding of the legal and practical environment of contemporary American cultural resource laws and regulations. The Toolkit is an essential resource for anyone working in the field and ideal for training archaeology students in classrooms and field schools.

Volume 1: *Archaeology by Design*
By Stephen L. Black and Kevin Jolly

Volume 2: *Archaeological Survey*
By James M. Collins and Brian Leigh Molyneaux

Volume 3: *Excavation*
By David L. Carmichael and Robert Lafferty

Volume 4: *Artifacts*
By Charles R. Ewen

Volume 5: *Archaeobiology*
By Kristin D. Sobolik

Volume 6: *Curating Archaeological Collections:*
   From the Field to the Repository
By Lynne P. Sullivan and S. Terry Childs

Volume 7: *Presenting the Past*
By Larry J. Zimmerman

ARCHAEOBIOLOGY

KRISTIN D. SOBOLIK

ARCHAEOLOGIST'S TOOLKIT
VOLUME 5

A Division of Rowman & Littlefield Publishers, Inc.
Walnut Creek • Lanham • New York • Oxford

Acknowledgments

I would like to thank Steve Bicknell, Rick Will, Tom Green, Larry Zimmerman, Kathryn Kamp, Tom Whittaker, Vaughn Bryant, Harry Shafer, Dinah Crader, David Sanger, Tom Alex, Larry Elrich, and Heather McInnis for their help in completing this book.

ALTAMIRA PRESS
A Division of Rowman & Littlefield Publishers, Inc.
1630 North Main Street, #367
Walnut Creek, CA 94596
www.altamirapress.com

Rowman & Littlefield Publishers, Inc.
A Member of the Rowman & Littlefield Publishing Group
4501 Forbes Boulevard, Suite 200
Lanham, MD 20706

PO Box 317
Oxford
OX2 9RU, UK

British Library Cataloguing in Publication Information Available

Library of Congress Cataloging-in-Publication Data

Sobolik, Kristin D. (Kristin Dee)
　　Archaeobiology / Kristin D. Sobolik.
　　　　p. cm. — (Archaeologist's toolkit)
　　　　ISBN 0-7591-0401-8 (cloth : alk. paper) — ISBN 0-7591-0023-3 (pbk. : alk.
　　paper)
　　　　1. Archaeology—Methodology. 2. Biology—Methodology. 3. Animal
　　remains (Archaeology) 4. Plant remains (Archaeology) 5. Paleoethnobotany.
　　6. Archaeology—Field work. I. Title. II. Series.

　　CC75.7 .S59 2003
　　930.1'028—dc21   2002015058

Printed in the United States of America

♾™ The paper used in this publication meets the minimum requirements of American National Standard for Information Sciences—Permanence of Paper for Printed Library Materials, ANSI/NISO Z39.48-1992.

 CONTENTS

SERIES EDITORS' FOREWORD

The Archaeologist's Toolkit is a series of books on how to plan, design, carry out, and use the results of archaeological research. The series contains seven books written by acknowledged experts in their fields. Each book is a self-contained treatment of an important element of modern archaeology. Therefore, each book can stand alone as a reference work for archaeologists in public agencies, private firms, and museums, as well as a textbook and guidebook for classrooms and field settings. The books function even better as a set, because they are integrated through cross-references and complementary subject matter.

Archaeology is a rapidly growing field, one that is no longer the exclusive province of academia. Today, archaeology is a part of daily life in both the public and private sectors. Thousands of archaeologists apply their knowledge and skills every day to understand the human past. Recent explosive growth in archaeology has heightened the need for clear and succinct guidance on professional practice. Therefore, this series supplies ready reference to the latest information on methods and techniques—the tools of the trade that serve as handy guides for longtime practitioners and essential resources for archaeologists in training.

Archaeologists help solve modern problems: They find, assess, recover, preserve, and interpret the evidence of the human past in light of public interest and in the face of multiple land use and development interests. Most of North American archaeology is devoted to cultural resource management (CRM), so the Archaeologist's Toolkit focuses on practical approaches to solving real problems in CRM and public archaeology. The books contain numerous case studies from all parts of the continent, illustrating the range and diversity of applications. The series emphasizes the importance of such realistic considerations as budgeting, scheduling, and team coordination. In addition, accountability to the public as well as to the profession is a common theme throughout the series.

Volume 1, *Archaeology by Design*, stresses the importance of research design in all phases and at all scales of archaeology. It shows how and why you should develop, apply, and refine research designs. Whether you are surveying quarter-acre cell tower sites or excavating stratified villages with millions of artifacts, your work will be more productive, efficient, and useful if you pay close and continuous attention to your research design.

Volume 2, *Archaeological Survey*, recognizes that most fieldwork in North America is devoted to survey: finding and evaluating archaeological resources. It covers prefield and field strategies to help you maximize the effectiveness and efficiency of archaeological survey. It shows how to choose appropriate strategies and methods ranging from landowner negotiations, surface reconnaissance, and shovel testing to geophysical survey, aerial photography, and report writing.

Volume 3, *Excavation*, covers the fundamentals of dirt archaeology in diverse settings, while emphasizing the importance of ethics during the controlled recovery—and destruction—of the archaeological record. This book shows how to select and apply excavation methods appropriate to specific needs and circumstances and how to maximize useful results while minimizing loss of data.

Volume 4, *Artifacts,* provides students as well as experienced archaeologists with useful guidance on preparing and analyzing artifacts. Both prehistoric- and historic-era artifacts are covered in detail. The discussion and case studies range from processing and cataloging through classification, data manipulation, and specialized analyses of a wide range of artifact forms.

Volume 5, *Archaeobiology*, covers the analysis and interpretation of biological remains from archaeological sites. The book shows how to recover, sample, analyze, and interpret the plant and animal remains most frequently excavated from archaeological sites in North America. Case studies from CRM and other archaeological research illustrate strategies for effective and meaningful use of biological data.

Volume 6, *Curating Archaeological Collections*, addresses a crucial but often ignored aspect of archaeology: proper care of the specimens and records generated in the field and the lab. This book covers strategies for effective short- and long-term collections management. Case studies illustrate the do's and don'ts that you need to know to make the best use of existing collections and to make your own work useful for others.

Volume 7, *Presenting the Past*, covers another area that has not received sufficient attention: communication of archaeology to a variety of audiences. Different tools are needed to present archaeology to other archaeologists, to sponsoring agencies, and to the interested public. This book shows how to choose the approaches and methods to take when presenting technical and nontechnical information through various means to various audiences.

Each of these books and the series as a whole are designed to be equally useful to practicing archaeologists and to archaeology students. Practicing archaeologists in CRM firms, agencies, academia, and museums will find the books useful as reference tools and as brush-up guides on current concerns and approaches. Instructors and students in field schools, lab classes, and short courses of various types will find the series valuable because of each book's practical orientation to problem solving.

As the series editors, we have enjoyed bringing these books together and working with the authors. We thank all of the authors— Steve Black, Dave Carmichael, Terry Childs, Jim Collins, Charlie Ewen, Kevin Jolly, Robert Lafferty, Brian Molyneaux, Kris Sobolik, and Lynne Sullivan—for their hard work and patience. We also offer sincere thanks to Mitch Allen of AltaMira Press and a special acknowledgment to Brian Fagan.

LARRY J. ZIMMERMAN
WILLIAM GREEN

1

INTRODUCTION

*A*rchaeobiology is the analysis and interpretation of biological remains from archaeological sites. In the most inclusive sense, archaeobiology refers to the remains of anything that was once living: animal bone, shell, scales, plants, seeds, pollen, phytoliths, charcoal, parasites, hair, and the list could go on. In this book, I use the term *archaeobiology* in a more limited sense to refer to the analysis of animal and plant remains. I do this to separate what I discuss in this book from the analysis of human skeletal material, usually referred to as *bioarchaeology*, and other specialized biological remains, such as scales, parasites, hair, and organic residue, which are not as frequently observed in archaeological deposits. I want to discuss how to recover, sample, analyze, and interpret the most frequently excavated biological remains from archaeological sites because they are what most archaeologists encounter on a day-to-day basis.

Archaeobiology is a relatively new term. Implicit in its definition is the integration of the analysis of plant and animal remains, a synthesis that until recently was rarely attempted or achieved. Looking at the definitions of the two major disciplines incorporated within archaeobiology, zooarchaeology and paleoethnobotany, reveals their inherently cohesive nature. *Zooarchaeology* is defined as the study of animal remains from archaeological sites to understand the relationship between humans and animals at the subsistence and environmental levels (Reitz and Wing 1999). *Archaeobotany* was first defined by Ford (1979) as the collection and identification of botanical remains from archaeological sites. This term is not used as frequently as *paleoethnobotany*, which is the study of the relationships between plants and humans in prehistoric times using the collection, identification, analysis, and interpretation of plant materials recovered from archaeological sites. The difference between the two words is that *archaeobotany* refers exclusively to the technical side of such research, whereas *paleoethnobotany* refers to the scientific and interpretive arena.

GOALS OF ARCHAEOBIOLOGY

Two main goals are central to archaeobiological research. The first is to analyze the relationships between humans and plants and animals and their effect(s) on each other. Interaction between humans and the environment (including plants and animals) does not proceed in one direction only; humans influenced the environment in as many ways as the environment may have constrained or provided opportunities for humans. For example, archaeobiologists analyze various results of agricultural and animal domestication, such as effects on human health, impacts on the environment and landscape, biological changes in plants and animals, and changes in human interactions and cultural systems associated with the development of domestication.

A case study of the analysis of environmental impacts due to agricultural practices was conducted by O'Hara and colleagues (1993) in central Mexico. The authors test the hypothesis that a return to traditional, prehistoric agricultural methods would be better for the environment than plow agriculture with draft animals, as used in historic and modern times. They collected twenty sediment cores from Lake Patzcuaro in the highland regions northwest of Mexico City to analyze soil erosion through time. Their data, which include pollen and diatom studies from other cores, indicate that three periods of extensive erosional rates in the past equal and surpass the erosional degradation that is taking place today in the region due to modern agricultural techniques. Based on this analysis of humans and the environment, the authors conclude that a return to more traditional agricultural techniques would not necessarily be better for the environment of the region.

In another example, Wagner (1996) examines how an agriculturally dependent prehistoric Fort Ancient community in Ohio adapted to short- and long-term food shortages. The Fort Ancient people were "consummate maize agriculturalists" (Wagner 1996:256) who followed a seasonally based subsistence round in which communities gathered in villages with permanent houses during the spring and summer, dispersing to winter hunting camps during the fall and winter. Plant and animal remains from these sites indicate that the people focused on a few resources, such as deer, elk, black bear, corn, bean, and chenopod. In times of stress or lack of preferred resources, the Fort Ancient people relied more heavily on a wider variety of resources that were kept in storage pits. Wagner analyzed the plant remains from seasonal storage pits to conclude that during all seasons, the same types of plants were stored, both for use as food and for seed crops. Pits were also buried and concealed when villages were abandoned and families dispersed to their winter camps. Food storage of diverse plants was the main coping mechanism used to avoid short- and long-term food shortages in this Fort Ancient community.

These research directions show there are complex relationships between humans and their environment. Archaeobiological analyses can help archaeologists understand these complexities and integrate a larger database into analyses and interpretations.

The second goal of archaeobiology is to place archaeologically derived information in its anthropological context. Archaeobiology is not a list of the types of charcoal found in a hearth or the kinds of animal bone identified from a shell midden. That information is meaningless unless it is applied to the interpretation of the activities of humans at the site or the environment in which humans lived. Cultural context is imperative when analyzing archaeobiological material. Without it, the research is nothing more than a list to be placed in an appendix to a report. Like artifacts, biological remains supply clues about lifestyles and cultural patterns of past peoples. Plants and animals were not just collected at random by a group of people; they were obtained and used for specific purposes.

For example, plant and animal products were used for fuel, and often specific criteria were used in their selection, such as availability, season, inherent heat value and combustibility, and social rules based on taboo or ritual. Archaeological and ethnographic records indicate that not all potential fuels in an area were actually chosen or used in all instances by a culture. To illustrate such a case study, in conjunction with two colleagues I analyzed plant and animal remains from twenty-eight fire pits from indoor and outdoor areas of a large pueblo in the Mimbres River Valley of southwestern New Mexico (Sobolik et al. 1997). The study was done to test the hypothesis that cooking was only conducted in outdoor fire pits and that indoor fire pits were mainly for warmth. Our analysis of bone, seeds, and charcoal from the fire pits indicates that, in fact, cooking was performed in both indoor and outdoor fire pits. However, hardwood fuel sources, such as oak, walnut, ash, cottonwood, willow, and boxelder, were chosen more frequently for indoor fires. Hardwoods produce a longer-burning fire with less smoke and would have been ideal for slow cooking methods or for longer-burning fires needed for warmth. Softwood fuel sources, such as juniper, pinyon pine, ponderosa pine, and Douglas fir, were chosen more frequently for outdoor fires. Softwoods tend to burn hotter and more quickly and produce more smoke because of their pitch content. Softwoods would have been ideal for quicker cooking methods, and burning softwoods in outdoor, open areas would be optimal considering the increased amount of smoke. Therefore, Mimbres people were selectively choosing particular wood types for different burning and cooking situations.

Another example of the usefulness of archaeobiological data to determine cultural patterns of prehistoric populations is in the analysis of seasonality of site or regional occupation. For example, Sanger (1996) tested two hypotheses on prehistoric settlement patterns in the Gulf of Maine. One hypothesis stated that prehistoric people resided along the coastal regions of Maine, particularly along river mouths during summer months, and moved to interior regions during the winter. Support for this hypothesis rests in ethnographic and historic documents that state that aboriginal peoples traded with Europeans during the summer as the former occupied coastal areas. Another hypothesis stated that prehistoric peoples moved extensively around the region and resided in coastal areas of Maine during all seasons. To test these hypotheses, Sanger used extensive seasonality data obtained through the analysis of growth rings from 874 softshell clams from sixteen coastal shell midden sites. The data indicate that coastal sites were occupied during all seasons and disprove the hypothesis that prehistoric peoples occupied the coastal areas only during the summer. The study also suggests an interesting modification of the second hypothesis: the possibility that year-round seasonality on the coast is evidence for two prehistoric populations, an interior group and a coastal group. Sanger's zooarchaeological study thus lays the groundwork for future research.

GOALS OF THIS BOOK

The purpose of this book is to discuss the best and easiest ways to retrieve, identify, sample, analyze, and interpret archaeobiological remains from archaeological sites. In essence, I provide a general cookbook of tried and true recipes, containing step-by-step directions on how to treat archaeobiological material. Just like recipes, these directions can be modified according to the type of site or material that you are analyzing. I am not suggesting that there is only one way to recover small fish bones from soil samples, quantify charcoal fragments, or interpret fiber remains. Instead, I present what I believe are the best and easiest ways to do these things, point you in the direction of other literature sources, and provide case studies from diverse areas and sites in North America to illustrate specific techniques or interpretations.

I also address a problem that has long been apparent in archaeobiological work: the inferiority complex. Archaeobiologists have complained (some loud, some long) that their work is not respected and tends to be placed in the appendix sections of archaeological reports or incorporated into other chapters in tabular form only. In truth, archaeobiologists are their own worst enemies. If they treat their data as an important cultural entity, then other archaeologists will treat it so, too. Archaeobiologists should be able to incorporate their data into "the big picture" and to reconstruct what their data reveal about prehistoric life. In many instances, archaeobiological analyses tend to be conducted by different researchers, and the results are reported separately. If an integration of the results is undertaken, it is the archaeologist, usually not a biological expert, who integrates these conclusions. Because such synthesizers are often unaware of the scope and limitations of each data set, their interpretations often lack potential insight or authority (Sobolik 1994).

In this volume, I emphasize the importance of archaeobiologists being involved in excavations from the beginning. In this way, they can help determine research design, how samples are collected, and which samples will be useful to help answer the research questions being asked. The more involved archaeobiologists are, whether in planning the research design or actually excavating the site, the more informed they will be when analyzing and interpreting the data and reconstructing the prehistoric picture. When archaeobiologists are involved in helping ask and answer questions and placing their data into a cultural framework, they won't be relegated to the periphery.

That is the ideal case. In the real world, especially in cultural resource management (CRM), research questions often aren't established or refined before a site is tested or excavated (see Toolkit, volume 1). Archaeobiologists often receive boxes of bone, bags of soil for pollen analysis, or flotation samples for seed and charcoal identification without having been a part of the original planning or excavation. The archaeobiologist at this point needs to obtain information on the site and how it was excavated and any contextual information the archaeologist can provide. The archaeobiologist may need to determine his or her own research questions that may or may not end up fitting with the overall site analysis and interpretation. Archaeobiological data and interpretations need to be more highly valued at all stages of research. By providing a good, sound data set and potentially innovative analyses and interpretations, archaeobiological data should be considered an integral component of the overall research design.

Ultimately, it is best to incorporate archaeobiologists in all stages of a research and/or CRM project. However, there are a number of stages in the recovery, analysis, and interpretation of archaeobiological remains that can be conducted by archaeologists who are trained specifically in such techniques. I provide a summary table to indicate which stages of a project can be conducted by trained archaeologists and which need to be conducted by archaeobiologists (table 1.1). In

Table 1.1. Stages of a Project in which Archaeobiologists Must Be Involved*		
Project Stages	Must Be Involved	Not Necessary Although Optimal
Project design		X
Analysis of taphonomy		X
Sampling design		X
Material recovery		X
Coarse screening		X
Fine screening		X
Sediment collection for flotation		X
Flotation		X
Sediment collection for pollen and phytoliths		X
Material analysis	X	
Quantification and analysis of plant remains	X	
Quantification and analysis of animal remains	X	
Pollen and phytolith processing and analysis	X	
Archaeobiological interpretation	X	

*It is always optimal, from a research project perspective, to include archaeobiologists at *all* stages of design, analysis, and interpretation to obtain the most synthetic and cohesive product.

sum, archaeobiologists should be involved in the research design process and in analysis and interpretation of plant and animal remains recovered from archaeological sites. Recovery of archaeobiological material can usually be conducted by trained archaeologists, although an understanding of potential problems and important techniques is essential. I summarize potential problems and important techniques in this book.

HISTORY OF RESEARCH

The New Archaeology of the 1960s, with its focus on cultural ecology, was significant in the development of archaeobiology. Archaeologists began to systematically save biological remains from archaeological sites as questions regarding diet, paleoenvironment, and ecology became important. With the recovery of more biological remains, due in part to methodological advances such as fine screening and flotation, came an increase in the quality of analyses and interpretation.

Although archaeobiology first became a significant focus for archaeologists as the New Archaeology was dawning, some excellent earlier studies of cultural patterns evidenced through biological remains set the stage for later systemic analyses. However, these studies tended not to be integrated with the data from other assemblages and were mainly conducted by specialists in the field. Paleoethnobotanists or botanists analyzed the plant remains from archaeological sites, and zooarchaeologists or zoologists analyzed the animal remains. These analyses usually were not integrated into the overall report, as is common with archaeobiological analyses conducted today.

PALEOETHNOBOTANY

Prior to the 1890s, the few paleoethnobotanical studies that were conducted were written mainly by botanists or people interested in natural history. This changed during the early 1900s when the emerging field of anthropology began training ethnologists to work with Native Americans on reservations and record what they supposed were the last bits of information still available about the Native American's past cultures and lifeways. When the early botanists studied North

American Indians, their purpose had been mostly utilitarian; all they wanted to record was information about plants and how those plants could be used in the present. On the other hand, early ethnologists collected several types of ethnobotanical data from the people they studied. The anthropologists' focus was on the Native Americans' point of view about the plants they used and how these plants fit into their view of the universe. Searching for a utilitarian value in the plants was not important to the early ethnologists.

A major advance in paleoethnobotanical work occurred at the Columbian Exhibition in Chicago in 1893. Part of the fair focused on the lives and artifacts of the North American Indian. Many exhibits showed different aspects of the Indians' life, including their uses of native plants. J. W. Harshberger examined dried plant materials from caves in Colorado for display at the fair. As a result of that study, he coined the term *ethnobotany* for this type of research (Harshberger 1896). After the fair, interest in ethnobotany among museums, government agencies, and universities increased. Two agencies that funded large numbers of ethnobotanical projects were the U.S. National Herbarium and the U.S. Department of Agriculture. Universities began ethnobotanical studies around the turn of the century, and the first Ph.D. in ethnobotany was awarded to David P. Barrows (1900) from the University of Chicago on the ethnobotany of the Coahuila Indians of Southern California. Barrows stressed that ethnobotanical studies must go beyond the applied or economic value of a plant and focus also on the role plants play in a group's social, religious, and folklore practices.

Melvin Gilmore's (1919) study of plant use by Plains Indian tribes was the first to note that even though some of these tribes were hunters and gatherers, their use of wild plants led to considerable modification of the environment. For example, he noted that groups often introduced plants to new regions. Through burning, Indians controlled certain weedy plants native to a particular region, and by encouraging other plants to grow, they increased the available quantity of desired plant products such as seeds or tubers.

More universities recognized the field of ethnobotany by the 1930s. In 1930, Edward Castetter established an M.A. program in ethnobotany within the Department of Biology at the University of New Mexico. Castetter and his students began to record the ethnobotany of the Indians living in the Southwest. In the late 1930s, R. E. Schultes established a program in ethnobotany at Harvard University with an emphasis on the search for new plants with medicinal merit.

Schultes and his students tended to focus mainly on the ethnobotany of Indians living in Central and South America.

In the mid-1930s, the University of Michigan created the Ethnobotanical Laboratory within the Museum of Anthropology. Gilmore and later Volney Jones headed this program, in which studies focused on plant remains from archaeological sites. In a lecture at the meeting of the American Association for the Advancement of Science in 1931, Gilmore (1932) described his research and requested that people save and send him plant remains from archaeological sites. Material from all over North America began to arrive at the lab for analysis. In most cases, he was permitted to keep the materials and send back only a report on what he had found. Gilmore believed the geographic influences and the physical environment must profoundly impact human habits and cultures. Unless the physical environment can be visualized, he believed cultural patterns could never be understood.

Gilmore's successor, Volney Jones, is often cited in archaeology and is considered the father of paleoethnobotany. Jones was head of the Ethnobotanical Laboratory for more than twenty-five years and analyzed a large number of botanical samples from sites in the eastern and midwestern United States during the 1940s and 1950s. The most significant site was Newt Kash Hollow, the first major paleoethnobotanical study conducted east of the Mississippi River (Jones 1936). In that study, Jones examined plant remains from Early Woodland deposits (later dated to ca. 700 B.P.) and reported at least eight native plants that he felt were cultivated or semicultivated. He was the first to report physical evidence of tobacco use in a prehistoric site east of the Rockies for the Early Woodland period, and he set the standard for explaining early Eastern Woodland subsistence patterns for years to come.

Other important questions then and now revolve around the origins of agriculture, which plants were domesticated, what was the character of the paleoenvironment, and how humans used their landscape. As plant remains from archaeological sites became important for testing hypotheses, other significant areas of research developed around methodological issues: What are the best techniques for recovering various plant parts from sites? How do we quantify the material that is recovered? How do we compare diverse data sets? Today, paleoethnobotanists deal with a large number of issues ranging from the technical aspects of recovery and identification to the interpretation of plant remains and their importance in broad-scale questions.

ZOOARCHAEOLOGY

Robison (1978) divided the history of zooarchaeology into three main time periods: Formative, Systematization, and Integration. The Formative period (ca. 1880–1950) saw the initial analyses of faunal material from archaeological sites. At that point, archaeologists were not systematically collecting faunal remains from sites, and the few analyses that were being conducted tended to be done by zoologists who were interested in the material for biological and environmental reconstruction rather than for archaeological purposes. Zooarchaeological research thus tended to be reported in biological publications. Archaeologists, when interested, tended to focus on one or two species, modified bone tools, or remains associated with human burials. Early studies tended to be descriptive in nature, although some studies foreshadow the types of questions and directions of study zooarchaeologists would take in the future. Such early work includes analysis of vertebrates and invertebrates from a Maine shell midden site that includes dietary hypotheses on the importance of different species based on their abundance (Loomis and Young 1912) and research on shells from an Arizona pueblo to determine trading routes (Fewkes 1896).

In the Systematization period (ca. 1950–1960) archaeologists started looking at faunal remains as a means of obtaining information on cultural behavior and adaptation, although theory and methods were starting to be designed and implemented. The most frequently cited article in zooarchaeological literature (White 1953) introduces the quantitative concept of minimum number of individuals (MNI), which is the most frequently used quantification method used on faunal material. MNI determines the minimum number of each species present at a site, and it can be analyzed according to the site as a whole or in relation to separate units, strata, or levels at a site. MNI is discussed in depth in chapter 4.

Lawrence (1957) urged analysts to change their focus from mere identification to interpretation so that meaningful and stimulating information could be obtained from faunal remains. During this period, the results of long-term, large-scale, integrative archaeological studies were being reported (e.g., Izumi and Sono 1963; Braidwood and Braidwood 1982), and the importance of faunal remains to archaeological interpretation was more widely recognized.

Zooarchaeology specialists started collecting and analyzing samples. Specialists included T. E. White, John Guilday, Paul Parmalee, and Stanley Olsen, who started to train students as zooarchaeologists. These specialists advanced zooarchaeological studies and allowed archaeologists to realize the amount of information that can be gained through the analysis of faunal material. The collections of faunal remains from excavations began to increase, and analyses started to appear in archaeological reports, although mainly as appendices. Zooarchaeology became a recognized and important field within archaeology.

All of these ideas came together during the Integration period, from the 1960s to the present, as the New Archaeology was promoted and as ecological approaches remain prevalent. Cultural ecology and environmental anthropology are the main frameworks of many analyses conducted today as zooarchaeologists integrate their research with other subfields of archaeology. Winters's (1969) analysis of faunal remains from sites of the Riverton Culture has been cited as the first significant zooarchaeological analysis of the New Archaeology era. Another important analysis was conducted by Smith (1975) on Mississipian adaptations in Missouri. He analyzed the remains from different site types in the uplands, lowlands, and swamps. He observed that the prehistoric peoples tended to have a base camp located on the ecotonal region between different microenvironmental areas. They would then exploit the different environments, depending on the season and the abundance of resources that area could provide.

The key issues addressed today by zooarchaeologists include (1) *taphonomy*, encompassing site formation processes, middle-range research, preservation and modification of site artifacts and ecofacts, and determination of cultural and noncultural site components; (2) *methodology*, entailing quantification, recovery, identification, and sampling; (3) *cultural ecology*, involving the relationship between humans and the environment, domestication of animals (which also has a strong biological component), subsistence strategies, human evolution, and human cultural lifeways; and (4) *biology*, encompassing paleoenvironmental reconstruction and ecology and morphology of various animal species.

THE ROLE OF CRM IN THE DEVELOPMENT AND FUTURE OF ARCHAEOBIOLOGY

The advent and development of CRM have taken archaeobiological analyses into the mainstream. CRM is a legal construct developed to assess and analyze cultural resources that may be impacted by economic and technological advances. Because plant and animal remains

1.1. CASE STUDY: ARCHAEOBIOLOGICAL INTEGRATION IN ANALYSIS OF CRM PHASE III EXCAVATIONS AT THE LITTLE OSSIPPEE NORTH SITE, MAINE

Archaeology at the Little Ossippee North site illustrates both a synthetic CRM Phase III project, which integrates archaeobiology at all stages of analysis (Will et al. 1996), as well as a contract-based project in which specific aspects can be used as research-focused projects (Asch Sidell 1999; Sobolik and Will 2000). Following Phase I and II study, the site was determined to be culturally significant and eligible for nomination to the National Register of Historic Places. Phase III excavations at the site were undertaken by Archaeological Research Consultants, Inc., as part of a federal dam relicensing application by Central Maine Power to mitigate anticipated erosion damage to the site over the term of the project license.

The Little Ossippee North site is situated at the confluence of the Little Ossippee and Saco Rivers in western Maine. Over one hundred square meters were excavated, representing less than 4 percent of the total site area. A large number and diversity of prehistoric remains were uncovered, including ceramics, lithics, and hearth features, as well as a diversity of archaeobiological remains, including animal bones, seeds, corn cobs, and charcoal. Geomorphological and soil analyses identified site formation processes, alluvial stratigraphy, fire history, and prehistoric flooding events. Three separate occupations were observed during excavation (figure 1.1): a Ceramic period occupation, ca. 1000 B.P. (horizon 2A); a Late Archaic period occupation, ca. 3000 B.P. (horizon 3A); and an Early Archaic period occupation, ca. 8000 B.P. (horizon 5A). Archaeobiological material was recovered from all three occupations, and analysis and interpretation of the remains were integrated into the discussion of each prehistoric occupation, not separated from site and context analysis and placed in separate chapters or in an appendix section.

During careful excavation, bone remains were recovered as a portion of large matrix samples containing many tiny fragments of bone. Bone was hand sorted from the matrix as well as recovered during coarse and fine screening. Plant remains were mainly recovered from flotation samples taken from features. Archaeobiological remains were preserved in the acidic soils because they were calcined and carbonized in addition to being covered with anaerobic alluvial deposits. These remains enhanced interpretations of the various cultural occupations at the site.

For example, the Ceramic period occupation contained many cultural features, which yielded 88.45 grams of carbonized plant remains. Flotation taken from the horizon 2A yielded an additional 174.47 grams of plant remains. Identification included an impressive list of maple, alder, birch, beech, pine, and oak wood and bark; acorn, beechnut, and hazelnut nutshells; and bunchberry, tick trefoil, strawberry, huckleberry, grass, cherry, raspberry, elderberry, blueberry, and grape seeds. The variety and amount of botanical remains recovered from the Ceramic occupation of the Little Ossippee North site is unmatched by any site in Maine. In addition, a large hearth was uncovered and 18.5 l of matrix was floated. The light fraction contained large amounts of carbonized wild plant food remains as well as two carbonized corn cob fragments. A radiocarbon date from hearth charcoal (1,010 ± 60 B.P.) indicates that these cob fragments may be the earliest evidence of domesticated plants in southern Maine. A recent AMS date on the actual corn itself returned a 570 ± 40 B.P. date, still the oldest corn in Maine (Richard Will, personal communication).

Analysis of 375 calcined turtle bone fragments from the horizon 2A in the southern portion of the Ceramic period occupation indicates that small, juvenile musk turtles were taken as food items, probably during a single, short-term procurement event. Size and age of the turtles provide another line of evidence, in conjunction with the wide variety of seeds, indicating the site was occupied during the fall. Adult turtles observed in the assemblage include the painted turtle and the wood turtle. These turtles were most likely obtained individually and in small numbers.

Analysis of the archaeobiological remains from the Little Ossippee North site was incorporated into additional research projects. The plant remains were included as a central database in Nancy Asch Sidell's (1999) analysis of the prehistoric subsistence patterns and paleoenvironment of Maine. She focused on the changing environment of Maine in the past and how subtle environmental differences would have affected human occupational patterns. She also presented evidence for the earliest agricultural practices in Maine, discussing where agriculture would have been feasible and how such practices would have affected prehistoric populations. Her analysis, in particular, illustrates that preservation of the potentially important botanical database is possible in Maine, particularly in alluvially deposited sites.

I analyzed the calcined turtle remains from the Little Ossippee North site. I was struck by the small size of the turtle bones, which led me to conduct an experimental project to assess the actual size of the turtles upon death (Sobolik and Will 2000). It is known that bone shrinks upon calcination; therefore, the turtle bones would have been larger and the turtles possibly older at death. The turtle-burning experiment involved burning modern turtle bone in an oven and in an open fire more similar to prehistoric conditions. The experiment revealed that turtle bone shrinks about 16 percent when it is burned. Therefore, 16 percent needs to be added to calcined turtle bone measurements to determine actual size of turtles at death. This addition still indicated that the turtles were small and juvenile. An analysis of turtle ecology suggested that the musk turtles were most likely captured as a group by prehistoric people when the juveniles were nestled in the muddy bottom of the slower-moving Little Ossippee River. A large quantity of small turtles would have made a more effective contribution to prehistoric diet than a few juvenile turtles caught singly over a longer period of time.

Analysis of the Little Ossippee North site illustrates CRM at its best. Archaeobiological recovery, analysis, and interpretation were essential to project goals and were incorporated at all phases of the project. Archaeobiological interpretations were included as an integral component in the discussion and write-up of cultural occupations; they were not relegated to a separate chapter or appendix with little or no integration. In addition, archaeobiological analyses and interpretations were used for other research projects that expanded from the original CRM project.

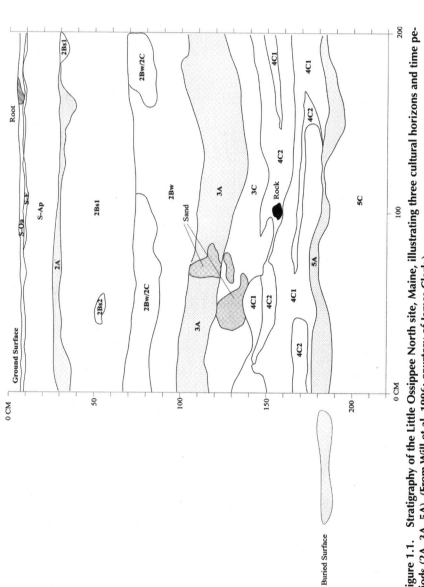

Figure 1.1. Stratigraphy of the Little Ossippee North site, Maine, illustrating three cultural horizons and time periods (2A, 3A, 5A). (From Will et al. 1996; courtesy of James Clark.)

are an integral part of cultural resources, their analysis is essential for any CRM project and report. Plant and animal remains are recovered during survey, testing, and excavation of potentially impacted archaeological sites. If plant and animal remains aren't recovered, assessments of the preservation potential of such remains are made.

Other archaeological projects, which may not be CRM driven, do not legally need to recover, analyze, and interpret archaeobiological material. Research-driven archaeological projects tend to focus on artifacts and remains that will help bolster or disprove a particular hypothesis. Artifacts and remains that are not necessary data for a particular hypothesis may not be recovered; if such artifacts and remains are recovered, they may not be analyzed. Artifacts and remains that are recovered but not necessary for hypothesis testing tend to sit in bags and boxes in laboratories or museums until they are needed as datasets for the testing of other hypotheses. In this scenario, such artifacts and remains (many of which are archaeobiological material) may never be analyzed.

Such is not the case for archaeobiological remains recovered as a part of CRM projects. As a case example, I review in sidebar 1.1 the goals, procedures, and results of a CRM project conducted by Archaeological Research Consultants, Inc., for Central Maine Power at the juncture of the Little Ossippee North and Saco Rivers. This case study not only illustrates the potential for recovery, analysis, and interpretation of archaeobiological remains as part of a CRM project but also highlights the importance of data recovered from CRM projects as tools to assess research questions and hypotheses regarding prehistoric lifeways in general. Archaeobiological analyses do not need to be limited to short discussions or appendices in CRM reports. Results from CRM-driven archaeobiological analyses may be an integral component for broader-ranging research questions that analyze prehistoric culture on a more global scale.

In addition, archaeobiological materials should always be recovered from research-driven archaeological projects. If they are not pertinent to the particular hypothesis or goals set forth by the researcher, then they should be properly curated and stored for future use. Archaeological remains are a nonrenewable resource, and excavation is entirely destructive. Therefore, recovery of archaeobiological remains from all archaeological sites is essential, not only for our present understanding of the past but also for the future's understanding of the past.

2

TAPHONOMY

Taphonomy is the study of site formation processes as they affect the preservation, inclusion, and distribution of biological components from archaeological sites. Efremov (1940), a paleontologist, defined the term *taphonomy* (*taphos*, tomb; and *logos*, law) as the study of the transition of animal remains from the biosphere to the lithosphere. Because the term was defined by a paleontologist, most taphonomic studies focus on recovery and analysis of bone remains, although taphonomy of plant remains is just as important for archaeological interpretations. Since its inception, the definition of taphonomy has been altered to fit the needs of both paleontology and archaeology. Today, the study of taphonomy has increased to include the post fossilization period (Lyman 1994), and the scope of taphonomy now includes the history of biological remains, including their collection and curation.

Because of the broad array of biological, environmental, and human agents affecting preservation, inclusion, and distribution of biological components in archaeological sites, taphonomy should be the first issue that all archaeologists think about before they even begin to recover these remains. Before walking onto a site, before screening for bones, before floating for seeds and charcoal, the archaeologist should be thinking about all the factors that could be responsible for the assemblage and could have affected the archaeological site overall. In this chapter, I discuss why it is important for the archaeologist to be aware of taphonomy at every step of investigation, what types of taphonomic factors could be influencing the assemblages, and how to account for these factors in overall analysis and interpretation.

"Taphonomic studies of modern analogs have shown the complexity of the processes that affect bones; but rather than despair, we should recognize that the processes likely to have operated at a particular archaeological site, and the likely range of variability in the patterned effects of those processes, are specifiable" (Bunn 1991:438). Every archaeologist should be able to specify which factors have influenced the assemblage he or she is analyzing and to incorporate that into site interpretations.

Archaeologists often determine taphonomy of a site or assemblage through the aid of experimental archaeology and ethnoarchaeology. Ethnoarchaeology and experimental archaeology provide methods for testing hypotheses about site formation processes and artifact assemblage preservation, movement, and origin. In experimental archaeology, scientists can conduct staged experiments as well as observe modern natural factors to examine the environmental and cultural elements that affect archaeological sites or assemblages. In this way, we can formulate ideas on how archaeological sites actually become formed and what types of impacts various factors have on the formation process.

For example, in a wood rat bone movement study, Hoffman and Hays (1987) controlled the experiment by introducing bone from six different animal taxa (dog, turtle, catfish, opossum, raccoon, and deer) into an active wood rat den in Tennessee to observe movement and modification of the bones by wood rats through time. The authors observed that six months after the bones had been introduced, 137 elements (46 percent) had been moved at least one meter; and of those moved, only 76 elements (55 percent) were recovered. The wood rat(s) had no preference for type of element, indicating that bone structure and morphology are unrelated to selection; the wood rat(s) moved bones as small as 0.3 gram (a turtle long bone) and as large as 101 grams (a deer pelvis). This research revealed that wood rats can be an important taphonomic factor in site formation and artifact movement, one that can be nonselective and can readily move cultural material from its primary context to a secondary location.

However, in a study by Bocek (1986), observation of modern rodent ecology was used to determine rodents' potential effects on archaeological sites. In that study, the experiment was not controlled; observation and interpretation of natural conditions were made. The ecology and burrowing behavior of a variety of rodents were studied to assess their possible effects on two central California archaeological sites in which rodent disturbance had been observed. In grassland environments, much of the soil surface has been disturbed by rodents, and burrowing behavior is a major soil formation factor. Rodent activity also tends to move soil and materials vertically, with smaller material accumulating near the surface and larger material moving to greater depths. The author estimates that archaeological sites have been "250% reworked by rodent activity" (p. 600) and that rodents have segregated the site into two stratigraphic horizons: an upper horizon of small materials and a lower horizon of large materials, which could easily be misinterpreted as cultural zones.

Another way in which archaeologists and archaeobiologists determine taphonomy of sites or assemblages is through the use of ethnoarchaeology. *Ethnoarchaeology* is the study of living human communities by archaeologists for the purposes of answering archaeologically derived questions. These studies are particularly useful to taphonomists because they document cultural processes involved in site formation. For example, in a study of modern Aché hunter-gatherer camps in the neotropical forests of Paraguay, Jones (1993) indicated that short-term camps exhibit a distinctive pattern that contrasts with long-term camps: The short-term camps have a fire-focused assemblage pattern in which debris is in primary context. Jones asked what types of patterns archaeologists should look for to identify short-term prehistoric camps. The author analyzed six one-night Aché camps that contained five to six fires per site for twenty-four to twenty-six people. Most of the activity at the camp took place within 1.5 meters of the fires, including butchering, cooking, and eating mainly small-sized animals. After the Aché left the camp, the main material remains were bones left in primary context; the short-term camps were not cleaned. The author suggests this pattern may be observed archaeologically when the group was foraging for immediate consumption at short-term camps and mainly used small animals. We can then infer that short-term camps of prehistoric people with a similar cultural pattern may contain artifacts in primary context, whereas long-term camps may exhibit more assemblages in secondary context.

There are limits to the use of ethnoarchaeological data because cultures of today have different behaviors and customs, and therefore different artifact and site patterns than cultures of the past. Because human behavior varies through time and across space, comparing the assemblage patterns of modern hunter-gatherers in Africa to Paleoindians in North America may be problematic. However, ethnoarchaeological studies do permit observation of cause-and-effect relationships between humans and their environment that cannot be obtained through other means.

In the rest of this chapter, I discuss factors that affect taphonomy of archaeobiological assemblages in a site. The preservation potential of each bone or seed is influenced by biological, environmental, and cultural factors (table 2.1). Of utmost importance is the determination of whether an archaeological assemblage is actually cultural (i.e., deposited or modified by humans). Biological remains can become deposited at archaeological sites in many ways that have nothing to do with humans. If you are interested in analyzing and interpreting past human behavior, it is important to ascertain which parts of your sites and assemblages are due to human behavior and which are not. After you determine which assemblages are cultural, then you need to assess whether that cultural material is in primary or secondary context. Is this cultural material in the context in which prehistoric humans placed it (primary), or has it been moved or modified by other processes (secondary) such as fluvial action, prehistoric dogs, rodents, tree throws, or pothunters? If you determine that your assemblage is

Table 2.1.   Relative Preservation Potential of Archaeobiological Materials

	Bone		Shell	Plant Remains	Pollen
Taphonomic Factors	Organic Component	Mineral Component			
Biological					
Saprophytic organisms	−	0	−	−	−
Material durability	+	+	+	+	+
Carbonization/ calcination	+	+	+	+	+/−
Environmental					
Acidic soils	+	−	−	+/−	+/−
Alkaline soils	−	+	+	−	−
Mechanical destruction	−	−	−	−	−
Weathering	−	−	−	−	−
Hot, arid conditions	+	+	+	+	+
Cold, arid conditions	+	+	+	+	+
Anaerobic	+	+	+	+	+
Cultural					
Processing	−	−	−	−	−
Cultural use	+/−	+/−	+/−	+/−	+/−

+ = postive preservation factor
− = negative preservation factor
0 = neutral preservation factor

in secondary context, you need to assess whether analysis of that assemblage will be useful to your overall research agenda and/or questions. Is the time and money that will be spent on recovery and analysis of these materials going to be worthwhile in terms of information return?

PRESERVATION

There are numerous reasons why some plant and animal remains are preserved at a site and some are not. Preservation is affected by biological, environmental, and cultural factors (table 2.1). In this section, I discuss some of these factors, but this discussion is not exhaustive. It is up to each archaeologist to analyze and experiment on the types of preservational factors that may have or may be influencing their own site. Each site is different and was deposited under diverse conditions that affect whether the site becomes preserved in the archaeological record at all and whether the biological remains at that site can be recovered.

BIOLOGICAL PRESERVATION FACTORS

Many biological factors influence the preservation potential of plant and animal remains at archaeological sites (table 2.1). The most important one is the presence of *saprophytic organisms*. Saprophytic organisms are plants and animals that live on dead matter and obtain all their nutrients (nitrogen compounds, potassium, phosphates, oxidation of carbohydrates) by breaking down organic matter. Saprophytic organisms can include larger scavengers and rodents, but the term mainly refers to small organisms such as earthworms, insects, fungi, bacteria, small arthropods, and microbes. These organisms cause the decay and decomposition of most organic material, including biological materials at archaeological sites. The environment(s) in which these organisms flourish greatly influences whether biological assemblages will be preserved (see "Environmental Preservation Factors," p. 23).

Other important factors affecting whether biological remains preserve are their robusticity, durability, and density. The more durable a bone or plant part is, the longer it will survive attempted decay by saprophytic organisms and chemical decomposition. Carbonization (burning) of plants and woods makes them more resistant to destruction. During the carbonization process, chemical constituents of wood and plants are converted to elemental carbon, a durable substance that offers no nutrients for saprophytic organisms. Therefore, in many regions of the world in which archaeological plant remains are usually degraded, charcoal (of wood or plants) may be recovered. However, complete burning or incineration can destroy charcoal, turning it to ash, which is of less archaeobiological value than charcoal.

Burned animal bone, representing various stages toward calcination, also will preserve better than unburned bone under certain conditions. Burning removes protein and alters the calcium content of bone. Calcined bone is pure white, friable, and porous, whereas bone that is not quite calcined (gray to white in color) is not as fragile and is quite strong. Calcined or almost-calcined bone preserves well in areas with acidic soils where unburned bone is degraded through chemical action (see "Environmental Preservation Factors," p. 23).

Plant remains that are most frequently preserved and recovered by archaeobiologists are those containing materials that have a structural or protective role for the plant and are therefore more durable. Such constituents include cellulose, sporopollenin (the main component of pollen), silica (the main component of most phytoliths), lignin, cutin, and suberin that are found in pits, seeds, rinds, spines, woody components, resin, pollen, and phytoliths. Plant parts that do not contain these durable elements will tend not to be preserved in archaeological sites, and their potential absence should be realized.

Bone, horn, antlers, teeth, hooves, hide, and shell are the most frequently observed animal remains at archaeological sites, again due to their durability. These materials resist decay in many settings because they are made of robust structural elements such as keratin and collagen (horn and hooves), phosphatics (bone, antler, teeth), chitin (insect and crustacean exoskeletons), and/or are calcareous (shell). A shell midden site may contain tons of durable, well-preserved oyster shell, and numerous Puebloan sites in the southwest contain large amounts of carbonized corn cobs, but this does not mean that prehistoric peoples were eating nothing but oysters or nothing but corn. Preservation differentially affects biological remains; some remains will be well preserved, and others will degrade or will become totally destroyed.

Bone is the most frequently observed type of animal remains from archaeological sites. All bone, however, is not created equal. Some bone is more resistant to decay and destruction due to its density. Different bone elements and bone elements from different species vary in structural density. For example, larger animals tend to have bone with greater density than medium and small animals, so their remains tend to be preferentially recovered from archaeological sites. An exception is beaver, whose dense bone has a greater durability than most carnivores and other medium-sized mammals. Denser elements include the jaw, femur, humerus, tibia, calcaneus, and astragalus. These elements will also be recovered with increased frequency over less dense elements. Mammal bone is denser than fish and bird bone and thus is more frequently recovered (Lyman 1984).

Another important and often underrated influence on the taphonomy of archaeobiological assemblages, particularly bone, is the dog. Dogs have been "man's best friend" for at least ten thousand to twelve thousand years, and their remains are found in numerous archaeological sites around the world. Unfortunately, dogs are destructive of bone assemblages. Hudson (1993) conducted an ethnoarchaeological study on the destructive effects that dogs can have on bone remains. She analyzed bone loss due to ingestion by domestic dogs among the Aka Pygmies, a modern group of hunter-gatherers living in the tropical forests of central Africa. The author documented hunting, butchering, meat redistribution, cooking, consumption, and discard at three residential camps. After the camps were abandoned, they were excavated, revealing that 74 to 97 percent of bone elements were lost, mainly from smaller taxa, due to domestic dog consumption. Due to their density, remains of large animals survive canine assaults best, as do skulls and limb shafts. The presence and effects of domestic dogs on bone assemblages can be ascertained not only through bone loss but through the presence of gnawed bones as well.

ENVIRONMENTAL PRESERVATION FACTORS

The environment in which plant or animal remains are deposited greatly influences whether they will be preserved for a month, a year, ten years, or one thousand years. Because saprophytic organisms are the main cause of biological material destruction, preservation mainly depends on what types of depositional environments are conducive or inhibiting to these organisms (table 2.1). Carbone and Keel (1985) listed four environmental factors that influence preservation of biological assemblages: soil acidity, aeration, relative humidity, and temperature. In addition, other geological conditions may also be important for biological preservation.

Saprophytic organisms are intolerant of highly acidic soils and live almost exclusively in alkaline soils. Therefore, acidic soils will tend to preserve organic components of biological materials because they are not destroyed by microorganisms. Alkaline soils will tend to have poor preservation of organic components of biological materials due to increased saprophytic activity. This can be seen in the potential for pollen preservation in the Southwest. Because soils in the Southwest are highly alkaline, preservation of organic materials in open areas tends to be rare. For example, Bryant and colleagues (1994) analyzed 509 pollen samples from soil collected by a CRM firm along a proposed pipeline route in deposits ranging in age from one thousand to five thousand years old. The deposits all had a pH value above 6.0 with anhydrous carbonates as the most common compounds. Only 243 (48 percent) of the samples contained a significant amount of fossil pollen; the rest were almost entirely devoid of fossil pollen. The mean pollen concentration values from the pollen bearing samples was 6,545 grains/gram of soil. In addition, 90 modern surface soil samples were collected in west Texas to use as control samples for the fossil pollen record. All 90 samples contained a significant amount of pollen with an average concentration value of 21,311 grains/gram of soil. The authors state that the high-alkaline environment of the Southwest is partly the cause of poor pollen preservation in fossil soil samples, and that pollen concentration values should be determined in this region to help the analyst judge the possible amount of pollen destruction in each particular soil sample. Pollen has an outer covering (exine) made of sporopollenin, which is one of the strongest natural substances known. However, alkaline soils are conducive to fungi and bacteria that eat pollen, creating a biased array in which pollen with more sporopollenin in its exine is better preserved than pollen with less.

Alkaline soils, however, tend to preserve mineral components better than acidic soils. Bone is made up of minerals (hydroxyapatite, calcium carbonate, trace elements) and organics (collagen, bone protein, fats, lipids) in an approximately 2:1 ratio. The organic components of bone will tend to be eaten by saprophytes in alkaline soils, leaving mineral bone components intact. Therefore, the mineral (structural) components of bone are preserved and can be recovered in alkaline conditions. Bone tends not to survive in acidic conditions because acids dissolve structural bases of minerals, leaving only organic traces of bone in the soil. Calcareous shell and antler are preserved in the same conditions as bone.

While alkaline soils tend to destroy organic assemblages (plant remains and organic components of bone) due to saprophytic activity, such soil types tend to preserve bone and shell (mineral components) better than acidic soils. For example, soils in Maine tend to be acidic, which limits the preservation potential of the mineral component of bone. When bone is preserved at interior sites, it has been calcined or almost calcined (see "Biological Preservation Factors," p. 21). Although only small pieces of calcined or almost-calcined bone are preserved in interior sites with acidic soils, bone (mainly uncalcined) is prevalent at archaeological shell midden sites along the coast. Bone preserves in these sites because weathering and degradation of the calcareous shell matrix produces an alkaline environment conducive to preservation of mineral components of bone. Therefore, bone preservation at archaeological sites in Maine depends on site location and soil alkalinity.

Even though alkaline soils, such as in a shell midden, tend to preserve bone, bone still decays, degrades, and alters through time, as was observed in an experimental archaeology study on bone from a shell midden on the northwest coast. The mineral fraction of bones from a shell midden located in San Juan Island National Park, Washington, was measured to test the hypothesis that bone decomposes in alkaline deposits. Linse (1992) analyzed salmon vertebrae from two different deposits (facies) in the shell midden, one with high pH values (8.4–8.8) and one with pH values near neutral (7.8–7.9). Bone mineral decomposition, measured as the calcium/phosphorus (Ca/P) ratio, was assessed using inductively coupled plasma spectrometry (ICP) on the archaeological salmon bone as well as a control sample of modern salmon bone. The author observed no significant difference between the two archaeologically derived bone assemblages, but the archaeological bone mineral decomposed as compared to modern bone. She concludes, "Bones may be relatively 'safe' in a shell midden in terms of traditional morphological classification; however, the chemical integrity of bone in alkaline environments is suspect" (p. 342).

Soil alkalinity is not the only factor that influences preservation of biological assemblages. Other factors include the mechanical destruction of biological remains through seasonal freeze/thaw cycles, increased biological preservation potential in dry/arid or dry/frozen regions, and anaerobic environments which decrease saprophytic activity. For example, although soils in Maine are acidic and tend not to preserve mineral components of bone, such acidic soils, because they inhibit saprophytic organisms, should preserve other organics like botanical remains. This is not the case, however, because of the seasonal freezing/thawing cycle that mechanically destroys chemical composition. Preservation of organics in this type of environment can be achieved when they have been carbonized or burned, making them more structurally durable.

An example of excellent preservation in dry/arid regions derives from the southwestern United States. As previously discussed, alkaline soils in this region are not conducive to pollen preservation. However, the Southwest is famous for preservation of other biological remains because of high temperatures and aridity and because saprophytic organisms do not thrive in hot, dry conditions. Preservation of organics in open areas of the Southwest, however, is not as good as in enclosed areas (caves, rock-shelters, pueblos, etc.) because of increased exposure such materials have to weathering (wind, erosion, rain).

Cold temperatures also limit the decay of biological remains because saprophytic organisms do not live in such an extreme, cold environment. Biological assemblages in the Arctic can be as well preserved as in dry, hot regions; however, most archaeological sites in the Arctic are surface sites with little deposition overlying assemblages. Therefore, some types of biological remains will weather extensively and become degraded over time. For example, large animal bone remains are ubiquitous at these northern sites, whereas botanical remains are harder to recover (recovery is also due to cultural impacts; see "Cultural Preservation Factors," p. 29).

Saprophytic organisms cannot live without oxygen, so anaerobic (lacking oxygen) environments will be conducive to biological preservation. Such environments include peat bogs, which are famous for preservation of "bog bodies" and other biological remains, and waterlogged sites, where wood is commonly preserved, such as stakes from prehistoric fish weirs. For example, numerous waterlogged pieces of wood have been recovered from the peat and silt beneath the city of Boston (Kaplan et al. 1990). Previous analyses indicate that the wood represents the remains of a number of prehistoric fish weirs used for trapping fish along the ancient shore. For this analysis, the authors examined 216 wood samples recently recovered from deposits dated to ca. four thousand to five thousand years ago. Wood taxa identified included beech, oak, alder, sassafras, hickory, maple, Canada hemlock, dogwood, birch, elm, ash, and bayberry. This assemblage indicates that prehistoric peoples collected wood from "upland and riparian habitats in fall and winter and probably constructed and repaired their weirs in the very early spring" (p. 527).

Relatively anaerobic conditions also exist under thick layers of clay or silt deposits, a good environment for preservation of biological materials. One reason carbonized botanical remains were preserved at the Little Ossippee North site in Maine (see the case study in chapter 1's sidebar) is that soon after humans made and used hearths at the site, a flooding event of the Saco and Little Ossippee Rivers capped the site with clay and silt deposits, preserving botanical remains in a relatively anaerobic environment. Depth of deposit of biological materials is an important component of preservation in such environmental conditions: The deeper material is buried, the more anaerobic the environment is, and the better the preservation potential.

We observed this principle when a nearby llama farm donated a llama skeleton to the zooarchaeology collection at the university on one condition: We dig up and collect the skeleton. The llama had been buried for three years in what the owners termed a "shallow" grave. One weekend, I took a group of graduate students to dig up the llama under the assumption that a llama buried for three years in a shallow grave in Maine's wet soils would be nothing but bones. We dug and we dug. The llama turned out to be deeply buried (1.5–2 meters) and was in very pristine condition with little decomposition of fur or muscle. We found a halter, and then we could tell why the llama was called Black Beauty. I told the owners we would be back in ten years. The deep burial was in a relatively anaerobic environment that inhibited saprophytic activity, thus slowing decay and decomposition.

Any disturbance to a relatively anaerobic environment can introduce oxygen and change the environment to one in which saprophytic organisms can thrive and cause decomposition. For example, after exposing the head portion of the llama we introduced oxygen and aerated the soil around the llama by digging up and moving the matrix. Because of this disturbance, the head portion of the llama will probably decompose at a quicker rate than the undisturbed hind portion that maintained its relatively anaerobic burial conditions. Therefore, human disturbance can influence whether soils will be more conducive to biological preservation. Similar disturbance factors include rodents, worms, and insects that dig burrows or pits, plant roots that grow into the ground, and tree throws that expose previously undisturbed, relatively anaerobic environments to the aerobic environment.

Tree throws are also significant because they mix up and disturb the cultural and noncultural components of a site (figure 2.1). Most forests have long histories of disturbance, including tree throws induced by storms, floods, and other factors. However, after a tree decays, there may be little surface evidence of the disturbance. Tree throws increase the decay potential of biological materials, and they move cultural and noncultural materials around. They may mix material from deposits of different ages, making it difficult to obtain correct radiocarbon dates or even to recognize stratigraphy.

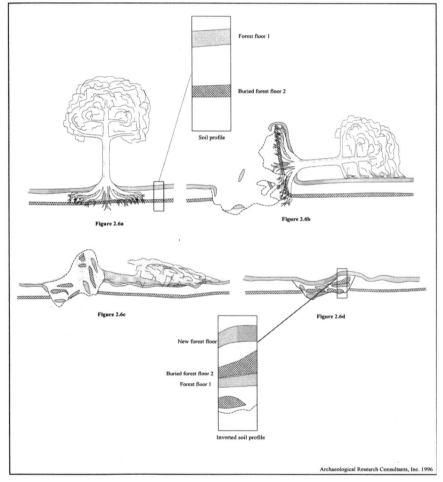

Figure 2.1.   Tree throw taphonomy as a means of inverting soil profiles and transporting older materials upward in the soil column. (From Will et al. 1996; courtesy of James Clark.)

CULTURAL PRESERVATION FACTORS

Humans also affect the preservation potential of biological materials (table 2.1). Before biological remains are deposited, people can affect their robusticity and structure, decreasing or increasing their preservation potential. People burn plants (carbonization) and bones (calcination), either intentionally or unintentionally, generally increasing the probability that those remains will be preserved (see "Biological Preservation Factors," p. 21). But humans also break, macerate, pound, chop, boil, and otherwise manipulate biological materials before deposition, decreasing their chance of preservation. For example, people break bone to gain access to marrow, sometimes pounding bones into small pieces and boiling them to extract the fat.

Humans also dig pits for various purposes, exposing underlying archaeobiological materials to a more aerobic environment, potentially reducing their preservation (see "Environmental Preservation Factors," p. 23). This type of cultural transformation is seen more frequently in large, multicomponent, or stratified sites where human activity was more extensive and diverse than it is at small, single-component sites where activity tended to be centralized and less invasive.

Ultimately, the main factor determining whether biological material will become deposited in archaeological sites is whether and how humans used that material or even brought it to the site. For example, gathering meat involves disarticulation and skinning of scavenged or hunted prey, and defleshing of bone. In the case of a large animal, much of this activity may take place outside the base camp; therefore, not all bones from the animal will be brought back to the base camp and become deposited to form the potential archaeological record. This was illustrated in an ethnoarchaeological study conducted by O'Connell and Hawkes (1988) in which they reported detailed observations on the food procurement practices of the Hadza, people of southern Africa, and the subsequent creation of bone assemblages. In general, they note that vertebra, scapulae, pelves, and upper limb bones are the most likely elements to be carried back to a base camp. The number of parts taken is a product of animal size, distance carried, and number of carriers. Elements that are the least damaged by the kill and disarticulation will be transported (axial parts, scapulae, and phalanges), while those most damaged (limb and cranial elements, and upper vertebrae) tend not to be transported. In addition, they observed that small animals are killed more frequently than large animals, but small animals are under-represented at kill sites due to total transport to base camps. When the Hadza killed larger prey, more bones were left at the kill site. Large animal butchery sites were also more complex, characterized by several clusters of debris and hearths, whereas small animal butchery sites tended to be composed of unstructured scatters of bone.

Another example of the effects humans have on preservation of biological remains and where differential parts are observed in the archaeological record is the processing of agave plants by peoples of the southern plains and southwestern deserts. Agave is a desert succulent with long, flat, sharp leaves aboveground and a compact "heart" or bulb belowground. The bulb is most nutritious just before the plant is ready to send up its reproductive stalk, so humans would dig up the plant at that time. First, the sharp leaves would be cut from the rest of the plant, and then the bulb would be dug up. The bulb would then be roasted in an earthen oven for at least forty-eight hours to make it edible (figure 2.2). People would eat agave at the earthen oven site or take it back to camp to share and eat there. Numerous chewed pieces of agave, called *quids* (figure 2.3), have been found in base camp sites. Remains of agave also will be found at procurement sites (leaves) and at and around earthen oven processing sites. In addition, people used agave leaves for basketry, sandals, paintbrushes, twining, and clothing, so fibrous remains of agave can probably be found in any site in which agave was used.

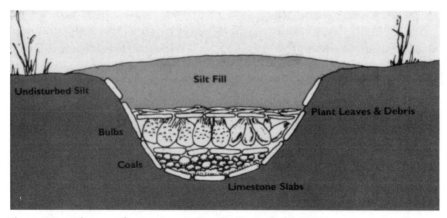

Figure 2.2.   Diagram of an agave roasting pit. (From Shafer 1986.)

Figure 2.3.   An agave quid. (Courtesy of Kristin D. Sobolik.)

CULTURAL VERSUS NONCULTURAL ASSEMBLAGES

When you recover biological remains, you need to assess whether the material is cultural. A *cultural context* is defined as a setting that has been physically altered or added to as a result of human activity. Evidence of human activity may include features (e.g., potholes, hearths, buildings), portions of stratigraphic layers (middens, living surfaces), or artifact concentrations. Noncultural contexts are contexts at which human activity is not indicated. Most sites have within them both cultural and noncultural contexts; that is, sites are commonly formed by a combination of cultural and noncultural processes. The archaeologist needs to determine which portion(s) of a site are cultural and which are not. The archaeobiologist can contribute greatly to this effort by determining which biological assemblages are cultural and which are not.

Most taphonomic studies addressing cultural versus noncultural deposits have focused on the accumulation and modification of faunal remains. These studies indicate that many animals, including carnivores, rodents, owls, and raptors, influence bone assemblages at archaeological sites. All of these animals collect and accumulate bones and may deposit bones into archaeological sites. Fortunately for the archaeologist, this noncultural bone generally is distinguishable from human-deposited bone. In addition, animals modify culturally used and deposited bone. This bone is still considered cultural, but it may be moved by animals out of original cultural (primary) context.

Carnivore and rodent influence on, or deposition of, a bone assemblage can be recognized by surface attributes of individual bone specimens. When chewing or gnawing, animals leave characteristic marks on bone. Microscopic examination of bone can reveal incisions, scratches, gouges, punctures, and pitting. Some of these marks are exclusively of human origin, while others are clearly of noncultural origin. Punched holes, striations, scoop marks, crunching, and splintering are examples of bone modification caused by animal activities. Canids will create shallow grooves or channels transverse to longitudinal axis on long bones because the long and thin shape of these bones prevents them from being gnawed in other directions. Punched holes, or tooth perforation marks, occur where hard bone is thin or where bone is soft, such as at the blade of the scapula or the ilium. These marks may appear as small hollows if the tooth did not fully pierce the bone surface. Striations and pitting occur on bone surfaces where canid molars scraped the surface in an attempt to reach the marrow cavity (figure 2.4). Tooth scratches tend to follow the surface of the bone, being deeper on convex surfaces and shallow on concave surfaces (figure 2.5). (In contrast, cut marks of human origin tend to be uniform in depth.) Where the bone ends have been removed, animal gnawing may produce scratches parallel or diagonal to the longitudinal axis of the bone on the shaft. Compact bone may be gnawed away to gain access to spongy bone, leaving overlapping striae and a scooped-out appearance on bone surfaces. Finally, marrow is reached by larger animals by crunching through bone, causing longitudinal splintering. Smaller canids will remove bone ends to weaken bone structure prior to crunching through the shaft.

Humans mark bone during butchering, skinning, and preparing food. Cuts are purposely placed for a desired result. For example, skinning an animal can leave cut marks on the underside of the chin and encircling the distal end of limb bones (figure 2.6). Because cultural marks are created on bone in butchering, cut marks cluster around articular surfaces or in areas of major muscle attachment. Marks will differ between species due to variation in joint strength. Bones struck with stone tools

Figure 2.4.   Canid gnawing, punctures, and striations on a long bone fragment. (Courtesy of Stephen Bicknell.)

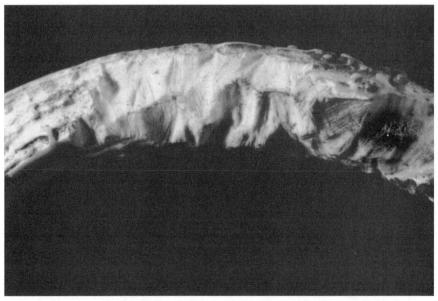

Figure 2.5.   Rodent gnaw marks on an antler. (Courtesy of Stephen Bicknell.)

will exhibit crescent-shaped notches at the point of impact, and bones broken during butchering by "grooving and snapping" show a heavy incision along the broken edge. Cut marks from tools are readily distinguishable from carnivore tooth damage. Tool marks are characterized by fine striations within the furrow made by cutting action. These striations are thought to be created by irregularities on the tool's working edge. Tooth marks lack striations, but they exhibit ridges perpendicular to the direction of the mark, caused by uneven force applied by the animal to the bone. These are often called *chatter marks*.

Archaeologists and archaeobiologists also have used the presence of small animals, particularly rodents, to distinguish cultural from noncultural bone, often considering small animal remains as noncultural. However, to disregard rodents and other small animals as possible human food sources underestimates the importance of small animals to the dietary array of prehistoric peoples. Bones of a wide variety of small animals have been recovered from paleofeces, indicating that they were eaten prehistorically, so their bone remains in archaeological sites may be due to cultural factors (Sobolik 1993). For example, numerous paleofeces from archaeological sites from southwest Texas contain bone remains from small animals; 333 paleofeces have been analyzed for their macrocontents, 245 (74 percent) of which contained small

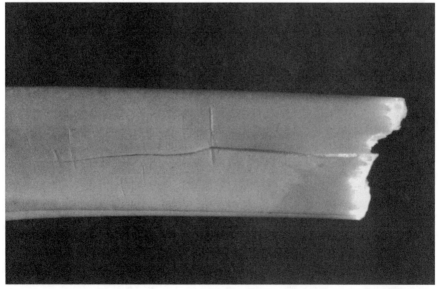

Figure 2.6.   Human stone-tool cut marks on long bone. (Courtesy of Stephen Bicknell.)

bone remains, and 123 (33 percent) specifically contained rodent remains (figure 2.7).

Differentiating cultural from noncultural bone deposits also requires analysis of the potential taphonomic factors that may have influenced site depositional processes and biological assemblage preservation. This analysis must be regional- and site-specific and fairly inclusive because factors may have operated in different areas and time periods. For example, I conducted a taphonomic study of the faunal remains from a prehistoric hunter-gatherer base camp in Big Bend National Park (see the case study in sidebar 2.1). Factors that influenced faunal deposition at the site were rodent burrowing and carnivore scat deposition. Potential taphonomic factors that turned out to be not important were fluvial deposition and modification and raptor pellet deposition.

Because of the wide variety of taphonomic factors that have influenced biological assemblages and the importance of understanding taphonomic history, archaeobiologists may become fixated on data collecting at the micro- or quantitative level without seeing the big picture. Before you examine five thousand bone fragments for the presence of cut, tooth, or gnaw marks, you need to assess whether such analysis is necessary for your overall research goals.

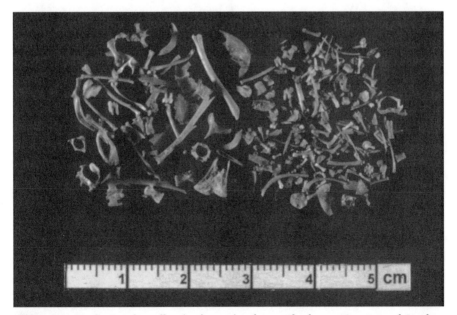

Figure 2.7.　Rodent and small animal remains from paleofeces. (Courtesy of Stephen Bicknell.)

2.1. CASE STUDY: TAPHONOMIC FACTORS AFFECTING ARCHAEOLOGICAL SITES IN BIG BEND NATIONAL PARK, TEXAS

I analyzed biological factors that may have influenced faunal assemblages deposited in a Late Prehistoric rock-shelter site in Big Bend National Park (41BS921). Cultural deposition at the site represents intermittent occupation by hunter-gatherers for approximately one thousand years. The animal bone excavated from the site included a variety of fairly well-preserved mammal, reptile, and bird remains, with a majority of the sample consisting of medium and large mammal bones (table 2.2; figure 2.8). During excavation, eight rodent burrows were discerned. The contents of these burrows were analyzed separately and revealed that the frequencies of bone classes were similar to those observed in supposed cultural levels of the site (table 2.2; figure 2.9). Bones from the rodent burrows and other areas of the archaeological site also exhibited similar frequencies of burning and weather erosion. These data indicate that rodents tend to move cultural material out of primary context but incorporate little noncultural material.

In addition, twenty-seven carnivore feces (most likely from mountain lion) were collected from the surface of the site and analyzed for bone content (table 2.2). Bone from these samples included mammal, reptile, and bird and were very fragmented and acid etched. Carnivore digestive tracts have strong acids to process food. When bone passes through this digestive tract it becomes dissolved or if preserved then extremely acid-etched. Two small areas of supposed cultural deposition at the site also contain such material, indicating the presence of degraded carnivore coprolites.

I also analyzed five great-horned owl pellets and excavated three large areas of owl pellet remains from Owl Cave, located one hundred meters from the site. The pellets and excavated faunal remains revealed a wide variety of animal bones from packrats, mice, rats, rabbits, snakes, and birds (table 2.2); the bone was in pristine condition with relatively few breaks (figure 2.10). Pristine bone was not observed in supposed cultural contexts from the archaeological site, indicating that owl pellet deposition probably did not contribute to the assemblage.

This study indicates that rodent burrowing at the archaeological site influenced cultural deposition and moved remains from a primary to a secondary context. In addition, carnivore scat deposition was also important in adding noncultural bone to the faunal assemblage, whereas owl pellet deposition was not a contributing factor in taphonomy of the site. This case study also illustrates the importance of determining the taphonomy of archaeobiological materials from supposedly cultural contexts.

Figure 2.8. Bone from a prehistoric rock-shelter in Big Bend National Park, Texas. (Courtesy of Stephen Bicknell.)

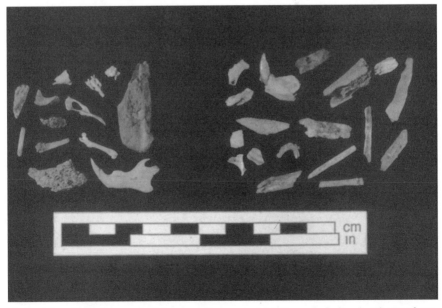

Figure 2.9. Bone from rodent burrows from a prehistoric rock-shelter in Big Bend National Park. (Courtesy of Stephen Bicknell.)

Figure 2.10.   Bone from great horned owl pellets. (Courtesy of Stephen Bicknell.)

What are your research questions? Why are you doing this? What is the big picture?

Here's an example of microlevel analysis addressing a big research design. Shipman (1986) examined a large number of bones under the scanning electron microscope looking for human and nonhuman marks from early hominid sites. She observed numerous instances in which human cut marks were superimposed on scavenger tooth marks, allowing her to conclude that early hominids were actually scavengers instead of hunters, obtaining meat after other carnivores had made the kill. Shipman's painstaking analysis was for a purpose: It contributed to the big picture. Even though all of our analyses may not have such far-reaching implications as Shipman's, we need to constantly be thinking of how our data fit into project-specific research designs as well as broader research goals (see Toolkit, volume 1).

CONTEXT

Ultimately, the question of whether a biological assemblage is cultural is a question of context. Interpretation of context occurs at all stages of research, from excavation to analysis. During excavation, the direct association of biological materials with artifacts or cultural features, such as stone tools, ceramics, or stone-lined hearths, provides evidence that the biological materials may have a cultural origin. When employing artifacts to make this determination, they should be clearly of human manufacture or modification and should be found in situ within undisturbed geological deposits that demonstrate primary association of artifacts with stratigraphy.

Disturbances should be evaluated critically. Potential disturbances are numerous and can include tree throws (figure 2.1) and carnivore and other animal modification as evidenced by presence of scat, burrows, and gnawing on bone (figures 2.4 and 2.5). Rodent burrows are common intrusions into archaeological contexts and are often easy to recognize during excavation. Rodents introduce noncultural material into archaeological sites and move cultural material out of primary context. These factors and many others can displace, introduce, or remove artifacts and archaeobiological materials from their original point of deposition. As previously discussed, humans, both prehistorically and in modern times, disturb archaeological context and move cultural materials from a primary to a secondary context. They may even introduce noncultural biological material into an archaeological context in which archaeologists will consider it cultural.

For example, Miller and Smart (1984) noted that animal dung is used as fuel throughout the world and that seeds can be found in dung. When dung is burned, charred seeds become incorporated into hearth deposits and might be misinterpreted as human food refuse. The authors analyzed four modern samples of debris from a village in Malyan, Iran: one hearth, one fire pit, and two midden samples that contained the remains of cleaned hearths and fire pits. Three of the samples contained charred seeds found only in dung and not directly used by humans. These results suggest that archaeologists should consider dung as a possible vector for charred seeds when the following conditions are met: (1) the site is located in an environment where wood and food are scarce, (2) dung-producing animals are present, (3) the archaeological assemblage includes parts of burned dung or seeds that are typically eaten by dung-producing animals, and (4) the archaeological context is one of hearths or dumping areas of cleaned hearths. The authors tested their research on two archaeological sites, a prehistoric Malyan site dated to the third millennium B.P. and the Tierra Blanca site, a late prehistoric site in Texas. Seed assemblages from both sites indicate that they may have been introduced through the burning of dung as fuel.

Biological material can also be disturbed and moved into or out of a site by water and wind. For example, moving water can remove biological remains from their primary archaeological context and redeposit them elsewhere. Typical secondary contexts are channels, floodplains, lake margins, point bars, and coastal settings. Many archaeological sites have been "discovered" that, in fact, are nothing more than artifacts in secondary context. Will and Clark (1996) conducted an experimental study in which they "planted" stone artifacts in the fluctuation zone on the shores of impounded Moosehead Lake in Maine to observe artifact movement patterns due to wave action and ice movement. After recording artifact movement over a year, they observed four main results: (1) Artifacts exposed to wave and ice action in the fluctuation zone can move several meters from their primary context; (2) the direction of movement is influenced by major weather; (3) artifacts moved by ice action have less surficial damage than artifacts moved by water and may look "fresh" or recently eroded; and (4) artifacts buried at the edge of the fluctuation zone in a beach or berm can be recycled to the surface by wave and ice action. These results indicate that a cluster of artifacts in the fluctuation zone or in the beach/berm area may represent material significantly removed from its primary context rather than material eroding from an archaeological site along the shoreline. These results also helped explain why many sites along a large impounded lake turned out upon excavation to be nothing more than surficial lithic debris. Although the experiment was conducted on lithic artifacts, the same processes can affect biological materials. For example, huge amounts of bone accumulate on sand bars in rivers across North America. These bone accumulations are not archaeological sites but merely water deposited assemblages in secondary context.

Fluvial action can move cultural remains out of a site and can also deposit noncultural materials into a site. Fluvial effects on bone have been extensively studied, whereas such effects on botanical material are not as well known. Surface abrasion and rounding of bone surfaces result from transport by water. Orientation may also be a sign that bone specimens have been moved by fluvial processes; heavier ends of elongate elements point upstream. Elements with low density (see "Biological Preservation Factors," p. 21), low weight, and a high surface area to volume ratio, such as innominates, scapulae, and intact crania, are more likely to be transported long distances. Shape will also influence transportability: Long flat bones are more likely to be transported than round ones.

In a classic experimental study on the effects of fluvial action on bone, Voorhies (1969) demonstrated this using bones from medium-sized animals in flume experiments and created a chart of elements and their transportability in flowing water (table 2.3). Group I elements were immediately moved by slow moving currents. Group II elements were gradually carried away in a moderate current, and Group III elements were only moved by strong currents. If your assemblage is composed of elements representing all of the groups, then it probably wasn't modified by fluvial action. If, however, you have an assemblage consisting only of elements from one group, fluvial action should be considered as contributing to the composition of the assemblage.

CONCLUSION

Determining taphonomy of biological materials recovered from an archaeological site is the first and most important step in archaeobiological analysis and interpretation. First, ask how the botanical or faunal remains became deposited in the site, and consider all the potential factors influencing that deposition. Are the biological remains from the site cultural, or do they represent deposition through noncultural agents? If remains are determined to be noncultural, they may be useful in analyses of paleoenvironments but not for direct analysis of human activity. Even if biological materials are considered to be deposited as a direct result of cultural activity, they may be out of primary context due to postdepositional factors, such as pot hunting, other human digging, animal burrowing, or tree throws. Depending on research goals, biological materials in secondary context may or may not be useful for analysis and interpretation; even though they are cultural, it may not be worth the time, money, and effort spent on their analysis.

In all stages of an archaeological project, decisions need to be made on the effectiveness and potential of each step. It is up to the archaeologist, in conjunction with the archaeobiologist, to determine which sites are to be tested or analyzed further in Phases II and III of CRM projects, and it is up to the archaeologist to determine which sites to collect data from and which will yield the most information during their short field research season. In most cases, the archaeologist is under time and money constraints. After the archaeologist, in conjunction with the archaeobiologist, has made these decisions and the site has been excavated, it is up to the archaeobiologist to determine

Context	Large Mammal	Medium Mammal	Small Mammal	Reptile	Bird	UID	Total
41BS921	593 (38%)	736 (47%)	129 (8%)	34 (2%)	9 (0.6%)	75 (5%)	1,676
Rodent burrows	7 (17%)	24 (57%)	1 (3%)	3 (7%)	0	7 (17%)	42
Owl pellets	0	2 (3%)	41 (54%)	32 (42%)	0	0	75
Owl cave floor	0	8 (2%)	177 (47%)	53 (41%)	5 (1%)	30 (8%)	273
Carnivore feces		234 (63%)		18 (5%)	10 (3%)	109 (29%)	371

Table 2.2. Bone from Taphonomic Contexts in Big Bend National Park

UID = Unidentifiable.

Table 2.3. Transportability of Bone Elements in Flowing Water

Easily Transported, Moved by Slow Currents	Intermediate between Easily and Moderately Transported	Moderately Transported, Moved by Moderate Currents	Intermediate between Moderately Transported and Transported with Difficulty	Transported with Difficulty
Group I Elements:	Group I/II Elements:	Group II Elements:	Group II/III Element:	Group III Elements:
Ribs	Scapula	Femur	Mandibular Ramus	Skull
Vertebrae	Phalanges	Tibia		Mandible
Sacrum	Ulna	Humerus		
Sternum		Metapodia		
		Pelvis		

Source: Data from Voorhies (1969).

which biological materials from these sites are worth spending diminishing supplies of time and money on. Which biological materials will help answer research goals, which materials will help in interpretation of past human lifeways, which materials are cultural and in context and which are not? Understanding taphonomy will help you answer these questions. Archaeobiological analyses and interpretation cannot proceed without this understanding.

3

 RECOVERY TECHNIQUES

Many methods exist to recover archaeobiological material from a site. Choosing appropriate methods depends on the type of site, the type of matrix and strata from which the remains will be recovered, and the kinds of research questions that are being asked. Presented here are what I consider to be the best and easiest ways to recover archaeobiological materials from most types of archaeological sites. I will also describe modifications that may be necessary depending on the type of site or matrix, emphasizing that modifications of the basic procedures for archaeobiological recovery are in the hands of project directors and archaeobiologists and depend on each individual site situation.

The most important principle in the recovery of archaeobiological material is that optimally the archaeobiologist should be involved from the very beginning (table 1.1). The archaeobiologist should participate in developing the research design and should help plan where excavations will take place and how they should proceed in reference to the recovery of archaeobiological material. During excavations, it is not necessary that the archaeobiologist be there to collect every biological sample. However, if the archaeobiologist is at the site, he or she can observe depositional conditions and recovery. In this way, the archaeobiologist will be able to assess taphonomic factors and conditions and will be able to modify the basic recovery plan if needed.

The least innovative and productive way to recover archaeobiological material is to have a field technician collect random, unspecified samples from the field and ship them to the archaeobiologist. The archaeobiologist has no indication of potential taphonomic factors; what cultural or environmental conditions may have influenced deposition and preservation; whether samples were collected from unambiguous cultural features, horizons, or zones; whether the "best" samples were collected; and how the samples relate to the entire site. If an archaeobiologist receives a bag of soil, a bag of seeds and charcoal, or a box of bones and has never been out to the site or been a part of excavation design and sample recovery, "analysis" may become nothing more than technical identification with little or no interpretation or innovation. If the archaeologist just wants a list of the types of charcoal or animal taxa present in a sample, then it is not necessary to treat the archaeobiologist as a scientist who can actually contribute to the analysis and interpretation of the site. And if the archaeobiologist wants to be treated as purely a technician rather than a scientist, he or she can keep accepting boxes of bone and bags of soil, along with a paycheck. If, however, the archaeobiologist would like to contribute to the understanding of peoples and environments and recognition of the importance of archaeobiological material in the reconstruction of past lifeways, then he or she should get involved at all stages of the recovery process.

Archaeobiology is becoming more completely integrated into archaeology. Accordingly, other volumes in this series cover certain elements relevant to archaeobiology. In particular, volume 1 discusses research design in detail, and volume 3 covers excavation methods. Therefore, I will limit my discussion of the recovery of archaeobiological material to the actual excavation of plant and animal remains.

Archaeobiological materials can be recovered in the same fashion as other archaeological material: During normal excavation and screening, some archaeobiological remains can be removed from the matrix or picked up in the screen. Some may be found during piece-plot excavations. However, because archaeobiological material often is too small to be observed during excavation and too small to be caught in normal screens, a lot of archaeobiological material is recovered during fine screening and flotation. Archaeobiological materials that are collected during normal excavation procedures should be bagged separately from those recovered during fine screening and flotation.

FINE SCREENING AND FLOTATION

Numerous studies show that using fine-mesh screens (one-eighth or one-sixteenth inch) and flotation is essential for the effective recovery of archaeobiological materials. Thomas (1969) provides baseline information on the recovery of animal bone from quarter-, eighth-, and sixteenth-inch mesh screens from prehistoric sites in the Great Basin. Bones from small mammals, weighing less than one hundred grams, were entirely recovered in the sixteenth-inch screens, but up to 54 percent of small mammal bones passed through the eighth-inch screen, and up to 93 percent were lost through the quarter-inch screen. Clearly, use of smaller-mesh screens greatly improves recovery of all animal taxa, particularly smaller taxa.

Shaffer (1992) also analyzed loss and recovery of faunal elements from quarter-inch screens. He placed modern, whole bone elements from twenty-five taxa, ranging in size from the least shrew (femur measuring 7 millimeters) to the coyote (femur measuring 176 millimeters), in a quarter-inch screen and recorded which elements passed through the screen and which were recovered. Predictably, smaller animal elements consistently passed through quarter-inch screen; rarely were elements recovered even for presence/absence determination. Some elements from larger animals consistently passed through the screen as well, such as caudal vertebrae, ribs, sternae, patellae, sesamoids, podials, metapodials, and phalanges from medium-sized animals, and sesamoids, carpals, patellae, and middle and distal phalanges of large animals. The author states that using only quarter-inch screening at an archaeological site will severely bias the sample toward larger animals and specific bone elements, thus making quantification unreliable.

The importance of flotation for the recovery of botanical remains has been recognized for many years; the study by Struever (1968) is usually cited as the first thorough North American discussion. Flotation uses water to separate lighter (less dense) material, usually organics, from heavier (more dense) material, usually inorganic matrix but also including some bone. Although flotation was used sporadically prior to the 1960s, the development of the New Archaeology, with its emphasis on ecological and environmental contexts of archaeological sites, led to increasing acceptance of flotation as the standard tool for botanical data recovery through the 1970s and its continued use today.

So, we understand that it is important to fine-screen and float sediments to recover archaeobiological materials. What are the best and easiest ways to do this? Fine screening is very simple: A known quantity of matrix is passed through a fine-mesh screen (usually eighth- or sixteenth-inch mesh), and all recognizable archaeobiological material retained in the screen is collected. Screening can be done by using a single screen (figure 3.1), inserted into coarse-screen frames (figure 3.2), or by placing the screen within or underneath the framework of an existing coarse screen (half-, quarter-, or eighth-inch), thereby screening the same matrix with coarse and fine screen. Nested screening, and dry screening in general, is easiest when dealing with dry and sandy or silty matrix. Wet or clay matrix usually is most effectively screened by water screening so that the matrix can be easily broken apart and the archaeobiological remains become more visible. Water screening involves placing a known quantity of matrix on a fine screen and using a pressurized source of water, such as a hose, to break apart the matrix and wash mud through the screen. It is important not to use too much pressure when water screening, or the archaeobiological remains will become smashed and broken and will pass through the screen. Fine screens can be purchased at any archaeological supply store or business, or they can be handmade with wood and screen purchased at the local hardware store.

Figure 3.1.   Example of a separate fine-screen system. (Courtesy of Kristin D. Sobolik.)

Figure 3.2.   Example of a nested fine-screen system. (Courtesy of Stephen Bicknell.)

The amount of matrix that should be fine screened and floated depends on the type of site being excavated and the project goals. It is important to obtain a representative sample of small archaeobiological remains from a site. To do this, you need to fine-screen and float matrix from features as well as from natural and cultural levels. The amount of matrix and sampling strategies for collection from five archaeological "type" sites are discussed in the section on "Excavation and Sampling," p. 53.

The advantage of fine screening and water screening is the increased recovery of small archaeobiological remains. The main disadvantage is that fragile remains, such as seeds, charcoal, and small bones, are easily broken during screening and by the impact of high-pressure water on the sample. Because of this problem, and because important material falls through even sixteenth-inch screen, flotation is usually the method of choice to recover fragile botanical remains. There are many flotation methods, and each archaeobiologist tends to prefer a specific technique. For a review of the types of flotation methods that have been employed by various projects around the world, see Pearsall (1989, 2000). I will review some of the techniques here and will then present what I consider to be the easiest technique with the fewest disadvantages.

REVIEW OF FLOTATION TECHNIQUES

The initial flotation method used by Struever (1968) was called the *immersion method* or the *apple creek method* and is useful in areas where there is ample water in which to immerse the samples. This method requires wash tubs with the bottoms removed and screens welded in their place. Partly immerse the tub in the water source (creek, river, pond, lake, etc.), then place the soil sample into the tub. Skim off all floating material with cheesecloth. Using many people and tubs simultaneously allows great volumes of material to be floated rapidly. The heavy fraction (the material that sinks to the bottom of the tub) is dumped in a bucket of zinc chloride solution (1.9 specific gravity) and separated further; some material such as bone that sinks in water, floats in zinc chloride. However, use of zinc chloride is problematic: The zinc chloride must be strained through cheesecloth after each use to prevent contamination, the process makes bones and calcium carbonate foam due to the hydrochloric acid in the zinc chloride preparation, the compound is costly, and it irritates skin and eyes.

Helbaek (1969) developed a similar technique to float samples in areas in which standing water is scarce. Soil is dumped in a bucket full of water, the soil is stirred, and the top portion of the water is poured onto a fine-mesh screen to collect the light fraction. The heavy fraction at the bottom of the bucket is either discarded or screened and dried for later examination. The process is repeated with new water for each sample. This technique, also known as the *washover method* (Greig 1989), can be used indoors or outdoors. Helbaek liked to use carbon tetrachloride rather than water because it had a higher specific gravity (1.8), which increased organic material recovery in the light fraction. Carbon tetrachloride is costly, however, and we now know that it is carcinogenic. The method works well with water, although bones tend to accumulate in the heavy fraction.

One of the most frequently used flotation methods today is the *oil drum method* or a modification of it. This method is also called the *SMAP* (Shell Mound Archaeological Project) *flotation method*, first used by Watson (1976). It involves pumping water from a nearby source into the bottom of a fifty-five-gallon drum that has a screen-bottomed bucket inset at the top. The constant addition of water agitates the drum's contents and creates a steady overflow of water. Soil is dumped into the screen-bottomed bucket; the light fraction floats, and the heavy fraction is caught at the bottom of the bucket or tub. The light fraction overflows the drum through a sluice and into a fine-mesh screen (0.425 or 0.5 millimeter), and it is then dumped onto newspaper to dry and then be sorted. This technique can process a large amount of soil even when water is limited because water in the fifty-five-gallon drum can be reused from sample to sample.

We used a similar technique at the NAN Ranch, a Mimbres Puebloan site in southwest New Mexico. A fifty-five-gallon drum was filled with water from a hose (figure 3.3). The drum had a large hole near the bottom that was closed with a screwtop. This hole was used to drain water and the heavy fraction once it had been used a number of times. Water entered the drum through a ring aeration system, creating a frothing effect that churned the soil and induced organics to rise to the top. Soil was dumped into the drum; the aeration system helped the light fraction float to the surface, where it would overflow the drum and descend a slanted spout attached to the top edge of the drum and be retained in cheesecloth set on a catchment area. The light fraction in the cheesecloth would then be hung to dry on a clothesline before being sorted and analyzed (figure 3.3). The heavy fraction would sink to the bottom of the drum, where it would be periodically drained through the large hole in the bottom side.

Figure 3.3.   Example of a SMAP flotation system used at the NAN Ranch, under the direction of Harry J. Shafer. This photo reveals water source, discard area, and drying flotation samples hanging in the background. (Courtesy of Kristin D. Sobolik.)

A potential problem with this and other mechanical techniques is that because several samples are floated using the same water, cross-contamination between samples is possible—unless you can empty the fifty-five-gallon drum after every sample, which is costly and time-consuming and uses a lot of water). In addition, in an attempt to limit contamination, each sample is run or floated for longer periods of time to increase the chance of obtaining most or all archaeobiological remains from a sample, leaving less behind to potentially contaminate the next sample. This can make mechanical systems more time-consuming than others.

COMBINATION METHOD

The method that I like to use is time-effective, has minimum potential for contamination, can be run by one person, uses small amounts of water, does not use nasty chemicals, and can be adjusted to individual soil sample types. In addition, it is also a fine-screening method so that one sample can be efficiently fine screened and floated at the same time. This method does not have a particular name but originates from the first basic manual flotation techniques generated before archaeobiologists started using more elaborate procedures. This technique does not have the limitations of the more evolved methods and retains all of their positive characteristics. For descriptions of this technique and its modifications see Pearsall (1989:35–50).

It involves a plastic or metal bucket from which the bottom has been removed and a fine screen (eighth- or sixteenth-inch) attached in its place (figure 3.4). The bucket is set into a tub or directly in a sink (if flotation is being conducted in the lab), or outside in an area that will become very wet and muddy (if flotation is being conducted in the field). If flotation is conducted directly in the sink, it is best that the pipe system have a sediment trap, or the pipes will become clogged.

A hose or a faucet runs water into the bucket and in turn into the tub. A known quantity of soil is emptied slowly into the bucket. The light fraction will float to the surface where it is collected with a fishnet skimmer and dumped onto a labeled, paper-lined tray for drying. The light fraction is collected continuously until organics stop floating to the surface. Flotation is assisted as the operator

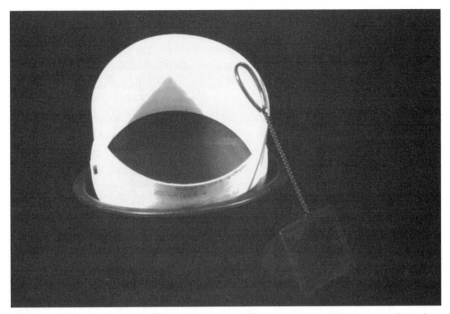

Figure 3.4. Example of combined flotation and fine-screen setup. (Courtesy of Stephen Bicknell.)

manually churns the soil to induce organics to float to the surface and soil to pass through the screen at the bottom. Each sample can be floated and manually assisted for as long as needed. After all of the light fraction has been collected, the bucket is removed from the tub, and the heavy fraction, which is automatically fine screened as well, is placed on a separate labeled tray for drying. The bucket and tub are then rinsed clean, and clean water is added to the system for the next sample, eliminating potential cross-contamination. This system can be operated by one person and can be run continuously, allowing numerous samples to be processed in a short time.

EXCAVATION AND SAMPLING

Even using the quick, easy, and efficient flotation and fine-screening method described here, fine screening and flotation are time-consuming and costly, so it is rarely feasible to run all or even a large quantity of the matrix from an archaeological site through such a system. However, in some cases, project goals are such that recovering all or most archaeobiological material preserved at a particular site is of primary interest. Such was the case at 20SA1034, a Late Woodland site in the Saginaw Valley of Michigan (Parker 1996), from which 665 flotation samples were recovered and processed from twenty-eight features, totaling over ninety thousand liters (see the case study in sidebar 3.1).

Which samples do you take for fine screening and flotation and from where do you take them? The answer to these questions depends on site taphonomy (discussed in chapter 2), research questions, and available resources. If you are interested in analyzing human diet and subsistence, you need to collect samples clearly associated with cultural areas of the site—again realizing that all portions of the site probably were not deposited or modified by humans. If you are interested in paleoenvironmental reconstruction, it is best to collect samples in areas that are not considered cultural. Paleoenvironmental sample collection therefore should take place away from the archaeological site so the information you are obtaining will have to do mainly with environment and not human selection. If you are interested in human impacts on environments and vice versa, then you should collect samples in both cultural and noncultural contexts.

Archaeobiological samples include both plant and animal remains collected from archaeological sites. As discussed previously, these samples should be collected both during normal coarse screening and with the aid of flotation and fine screening. I advocate the use of a combined fine screening and flotation system (see "Combination Method," p. 52), so in my discussions of where and how to take samples, I am referring to samples to be used for fine screening and flotation in combination. Archaeobiological samples should be collected with other artifacts during normal coarse screening and recorded and curated separately. Therefore, the following discussions refer to archaeobiological sample collection of fine-screen and flotation samples. In addition, I discuss collection of other archaeobiological samples, such as pollen, phytoliths, and paleofeces, where applicable.

Most archaeologists and archaeobiologists are interested in some aspect of human use of plants and animals. Archaeobiological samples should be collected in areas of a site that are considered cultural zones, horizons, or levels. In many cases, cultural affiliation is uncertain until artifacts and archaeobiological samples have been analyzed. In fact, archaeobiological samples can help determine whether an area of the site is cultural, whether rodents were rampant there, and

3.1. CASE STUDY: FLOTATION SAMPLING STRATEGY
AT A LATE WOODLAND SITE IN MICHIGAN

The recovery of evidence of horticultural practices during the Late Woodland period in Michigan is scarce. At first, archaeologists hypothesized that the scant evidence of horticultural botanical remains was due to lack of flotation in early excavations prior to 1975. Later, archaeologists believed that horticultural plants were not an important component of Late Woodland subsistence because even when flotation samples were taken, horticultural crops were not recovered in quantity. Excavation of site 20SA1034, a Late Woodland occupation in Michigan, illustrated that when 100 percent of cultural features are floated and analyzed, significant amounts of horticultural crops and other food remains can be recovered. Of particular importance is that not all flotation samples from the same feature yielded food plant remains. If the entire feature had not been floated, wood charcoal may have been the only botanical material recovered, and evidence for horticultural subsistence practices in the Late Woodland would have remained scarce.

Excavation of site 20SA1034 was sponsored by Great Lakes Gas Transmission Limited Partnership and conducted by the Institute for Minnesota Archaeology (Dobbs et al. 1993). The project research design involved maximizing recovery of archaeological remains from all twenty-eight cultural features of the low-density site; therefore, all feature matrices were processed for flotation, yielding 665 flotation samples and a total volume of approximately 90,440 liters (Parker 1996). Botanical remains recovered through flotation were analyzed according to a timeframe priority ranking in which food plant remains were identified first and then wood charcoal was analyzed. Wood charcoal was observed in all flotation samples, and a few features contained high concentrations of food plant remains.

At least fifty different plant food taxa were identified, the highest diversity recovered from that region. The plant remains included maize from twenty-four features with a density of 2.74 fragments/ten liters. Maize kernel and cupule remains indicate several different methods of processing as well as the presence of eight-row (Northern Flint or Eastern Eight Row) and twelve-row cobs. A diversity of nutshell fragments of acorn, butternut, hickory, and walnut was identified. A very large diversity of seeds also was identified (2,591), at an average of 2.9 seeds/ten liters. Sixty-three percent of the seeds were from cultivated or domesticated species such as sunflower, tobacco, chenopod, and cucurbit. In addition, a wide variety of wild plant seeds were also identified, such as black nightshade, huckleberry, sumac, elderberry, blackberry or raspberry, bunchberry, pin cherry, plum, and panic grass.

Parker concludes that if a standard ten-liter sampling strategy for feature flotation had been implemented, large categories of botanical remains would have been missed, such as the presence of sunflower and cucurbits. Through the successful application of total flotation and recovery of features a more accurate picture of Late Woodland subsistence in Michigan was revealed. Populations occupying the Saginaw Valley of Michigan during the Late Woodland time period practiced a mixed strategy of crop production and wild plant exploitation.

whether deposition represents carnivore habitation or alluvial deposition rather than or in addition to human occupation. Because cultural affiliation may be unknown or suspect, it is best to collect more samples rather than fewer samples.

As we all know, archaeology is a destructive science, and once a site has been excavated, it is gone. Therefore, it is better to collect more samples than you might actually analyze than collect fewer samples than you really need. Curation of unused and unanalyzed archaeobiological material is the necessary cost of doing competent, progressive archaeology in which needs of the future are assessed on the same levels as needs of the present. The future of archaeology may well rest on museum shelves as archaeological sites are potted, looted, destroyed, and excavated at an alarming rate. Therefore, the costs of curating archaeobiological remains in the present are minimal compared to their potential contribution to the future.

I present a basic procedure for sample collection (figure 3.1), a procedure that should be modified to fit the needs of each researcher and the circumstances of each site. In most archaeological sites, archaeobiological samples should be collected from sequential excavation levels so that a progression of samples is obtained from a unit. Samples do not have to be collected from every level of every excavation unit, but once an excavation unit is chosen for fine-screen and flotation sample collection, then samples should be collected from every level of that unit. Archaeological statisticians believe that for data to be useful statistically, samples should only be collected in a random fashion. If archaeobiological samples are collected randomly throughout the site, they lose context comparability, a very important analytic tool. In addition, samples chosen randomly would, in effect, be collected from many noncultural contexts that may be meaningless for project goals, and many important cultural contexts would be missed. Therefore, I believe that archaeobiological flotation and fine-screened samples should be taken from nonrandom, explicitly chosen contexts that fit with project goals.

Samples should be collected from every level within a chosen excavation unit to help in cultural zone determination and to be able to develop a progression of plant and animal use through time if the deposits represent cultural zones in chronological order. I usually collect samples from the southwestern quad of each level, but any quad is acceptable as long as collection protocols are consistent. Samples should have the same volume so that concentration comparisons can be made. A two-liter sample is usually of sufficient volume to obtain

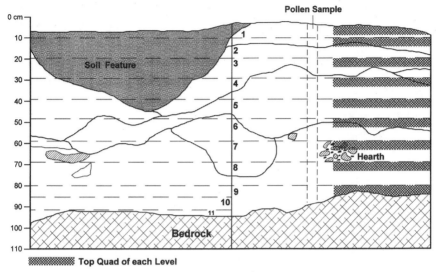

Figure 3.5. Example of basic flotation and fine-screen collection strategy in which two-liter samples are taken from the southwestern quad of each 1 × 1–meter excavation unit, as well as from sections of cultural zones or levels and cultural features. Pollen soil samples are taken from a column in the stratigraphy. (Courtesy of Stephen Bicknell.)

a representative sample from each level, but the volume collected should be increased if increased recovery is needed.

In addition to sample collection from each level, archaeobiological samples also should be taken from any feature or specific cultural context encountered, such as hearths, floors, caches, pits, and any other anomalies (table 3.1). Samples should be collected from these features even if they do not fall within units from which level samples are being collected. Most, if not all, of the matrix from small features should be collected for flotation. Large features should be sampled extensively, where possible, to take advantage of the heightened preservation potential that often exists in pits and other sealed deposits. For feature samples as well as level samples, sample volume should be carefully recorded.

The most important part of collecting samples from features is to include only matrix from the feature itself. Matrix from surrounding deposits should not be included in the sample. Therefore, excavation and sample collection surrounding and including features should proceed using natural rather than arbitrary stratigraphic control. If excavations are being conducted using arbitrary levels, excavation procedure should change to natural levels when features or cultural or natural strata are encountered, particularly if archaeobiological samples are to be taken in that area.

It is always important to record the volume collected and to catalog each sample within the project's cataloging system. Each bag is then taken to the field flotation center or back to the lab for subsequent flotation and fine screening. Bags used for collecting and recording other artifacts from the site can be used for collecting bulk samples for flotation and fine screening. In most cases, large plastic bags will be needed to hold all of the flotation and fine-screening samples in one bag. If plastic bags are used, flotation and fine screening should be undertaken relatively quickly, within a couple of days, or micro-organisms will start to flourish in the sample, potentially destroying organics. If flotation and fine screening are not going to take place relatively quickly, then soil samples should be collected in sandbags, which are slightly permeable, allowing slow drying.

Pollen and phytolith samples also should be taken throughout the site for analysis of human dietary patterns and from areas surrounding the site for paleoenvironmental analysis (table 3.1). Many researchers take pollen and phytolith samples for paleoenvironmental analysis from the archaeological site. But paleoenvironmental samples should be taken away from the site so that what is being analyzed reflects paleoenvironment and not human depositional patterns. In fact, paleoenvironmental reconstruction should proceed through the analysis of pollen and phytolith cores from lakes, bogs, or fens if such features are present. Pollen and phytoliths should be analyzed in conjunction with geomorphological analyses and with any other paleoenvironmental data that might be present.

Pollen and phytolith sample collection from archaeological sites involves collecting matrix samples from a column in an exposed profile (figure 3.5). Because the focus of sample collection in archaeological sites is understanding human activity, samples should be chosen from areas of the site that have good cultural context. In most cases, samples should not be collected randomly throughout the site. A column sampling technique from a fully excavated stratigraphic profile should be used so that potential changes through time can be delineated (see figure 3.5). In many cases, good cultural context is not be known until after excavation, so pollen column samples should be taken from a variety of areas throughout the site.

Pollen and phytolith column samples are taken after excavation has been completed in a unit and a stratigraphic profile has been exposed (Dimbleby 1985; Piperno 1988). First, you need to decide how

---

Table 3.1.   Standard Archaeobiological Sampling Procedure

Flotation and Fine-Screen Sampling

1. Select 1 × 1–meter units for sample collection. As excavation area grows, add units from which to sample.

2. Select one quad of the unit for sample collection, such as the southwest quad. Systematically collect a two-liter sample from each level of that quad using a bucket with the two-liter volume (or whatever volume you want to use) clearly delineated.

3. Collect samples from any archaeological features encountered, such as hearths, pits, caches, floors, and other anomalies, carefully recording sample volume. Samples should include only feature matrix, not surrounding matrix.

4. Collect samples from noncultural deposits for comparison with cultural deposits, carefully recording sample volume.

5. Assign each sample a separate, unique field sack or lot number.

Pollen and Phytolith Sampling

1. Select an exposed profile in good cultural context from a unit that has been entirely excavated.

2. Determine how many samples you are going to collect from the profile in column fashion. You can collect one or two samples from each cultural zone, from each stratigraphic level, or every five to ten centimeters along the column. Label bags according to the location of each potential sample and provide a separate field sack or log number for each.

3. Prepare for sample collection making sure all of your materials are handy. Collection needs to take place quickly and efficiently to avoid environmental pollen and phytolith contamination. Materials include a trowel, spoon or other small collecting tool, water, towel, and labeled baggies.

4. Start taking approximately one-hundred- to two-hundred-milliliter samples from the bottom of the column working to the top. The trowel and spoon should be cleaned with clean water and dried with a clean towel before each sample collection. Each sample should be taken after the profile has been scraped clean with a trowel. Take samples with a spoon and place each into a clean, labeled plastic bag. Rewash the trowel and spoon for the next sample collection.

5. Back at the laboratory the sample can be split for pollen and phytolith processing.

---

many samples you want to take from the profile. In most cases, you should take contiguous samples so that the samples are close together, and each will represent a range of pollen deposited from that time period. Samples are usually taken five to ten centimeters apart. Sometimes, however, samples can be taken strictly following the natural or cultural stratigraphy, particularly if the strata occur in levels thinner than ten centimeters. Mark where you want to take the samples with a measuring tape or flags beforehand.

Wind-blown pollen and phytoliths are ubiquitous in the natural environment, so contamination is an important issue. During sample collection, all tools and supplies must be cleaned and wiped to prevent airborne pollen and phytolith accumulation as well as contamination from one sample to the next. All supplies needed for sample collection should be at hand, and plastic collection bags should be prelabeled, each with its own field sack or lot number. Supplies needed include a trowel, spoon, water, cloth or towel, and prelabeled plastic bags. Sample collection should proceed from the bottom of the column up to avoid contamination from upper deposits. Use a clean and dry trowel to scrape the profile clean. At each designated sample collection spot, use a newly washed and dried spoon (or other useful collection device) to collect approximately one hundred to two hundred milliliters of newly exposed matrix and place it immediately into the appropriately labeled bag. Scrape the profile of the next sample collection spot with a newly cleaned and dried trowel, and use a newly washed and dried spoon to collect the sample. Because sampling damages the profile, this work should take place after excavation and profile mapping are complete. In addition to profile column sampling, pollen and phytoliths also can be collected from features such as pits, caches, hearths, burials, and floors.

To better understand whether pollen and phytoliths from an archaeological site are related to human activity or environmental conditions, you also should take samples off site. Such samples usually consist of modern matrix from the surface to compare with archaeological samples. For this collection method, you can either collect a number of one-hundred- to two-hundred-milliliter matrix samples, or you can use the "pinch" method and collect a number of pinches of matrix from around the site and combine them for a single modern pollen sample. Paleoenvironmental data can be obtained from pollen samples taken from a natural profile some distance from the archaeological site to use as a time/depth comparison to the archaeological samples. To do this effectively, both sample areas need to have good chronological control, usually through radiocarbon dating.

To illustrate archaeobiological sample collection techniques in different environments, I use examples from five different archaeological sites in the United States from which I have collected archaeobiological materials. Some of these examples represent actual sites, and some represent an amalgam of similar sites. The collection methods I advance are not necessarily the methods I used at each of these sites but in retrospect are those I would use if I could do it again. Again, I want to reiterate that each site is different, and the archaeo-biologist or archaeologist needs to make sample collection decisions based on the conditions of each site.

The first two site types I discuss are those in which most of the material represents prehistoric cultural deposition: a Puebloan site in the Southwest and a large shell midden site on the northeastern coast. Noncultural agents may have modified parts of the site, particularly postdeposition, but most of the material in the site itself consists of cultural debris. Site type 3 is a rock-shelter in which preservation of archaeobiological materials is excellent but taphonomic factors are numerous. Site type 4 is an open, ephemeral site with shallowly buried lithic scatters and evidence of possible postdepositional modifications such as plowing, erosion, or tree throws. Site type 5 is a deeply buried site with evidence of hearths and lithic scatters, along a river or lake that has flooded or been impounded. These site types do not represent all of the environments of archaeological sites in the United States, but they do cover a wide range of environments from which archaeobiological collection procedures can be extrapolated to similar environments and sites.

PUEBLOAN TYPE SITE

Type site 1 is a large Puebloan site in the Southwest in which excavations have been conducted in both the room blocks and in the midden and plaza areas. Preservation of archaeobiological remains in the arid Southwest is excellent, and large pieces of bone and charcoal are recovered during normal coarse screening. Fine-screen and flotation samples should be taken throughout the site because of the probability of obtaining well-preserved archaeobiological remains. Collection of samples in the midden should proceed according to the basic method I use to collect most archaeobiological fine-screen and flotation samples (figure 3.5; table 3.1).

Random 1 × 1–meter units should be chosen from which to systematically collect two-liter archaeobiological samples from the southwestern quad. As the excavation grows, the number of units chosen for sample collection should grow. I recommend, on average, that for every four to five 1 × 1–meter units chosen for excavation, one unit should be chosen for archaeobiological sample collection.

This average number will change depending on a variety of factors, such as time and money constraints, environmental conditions, and research questions.

Sample collection also should be done in the room blocks themselves. Room blocks tend to contain postdepositional fill (i.e., some secondary cultural deposits and some fill representing environmental noncultural deposits). Primary cultural deposits might not be encountered until excavation has proceeded to the bottom of the room. There the collapsed roof may be found on top of the room floor. Samples from deposits near and on the room floor should be taken so that prehistoric material is recovered. However, some postdepositional room fill may include midden deposits from later occupations, so it is important to obtain samples from these deposits as well.

Throughout the midden and in the room blocks, features may be found from which additional samples need to be taken. Features may include hearths, burials, pits, caches, and anomalous matrix. In these cases, the entire feature or a large portion of the feature can be collected for fine-screening and flotation. Again, it is important to record total volume amount, or two liter samples can be collected to stay consistent with other sample volumes.

Due to the arid environment, other types of archaeobiological samples can be preserved in the Southwest, such as fiber artifacts (basketry, sandals, netting), bone artifacts and tools, and paleofeces (also termed *coprolites*). Fiber artifacts and bone artifacts are easy to recognize during excavation and screening. Paleofeces are harder to recognize if you do not know what you are looking for. Paleofeces are desiccated human feces that contain the remains of what past peoples ate (figure 3.6). They can be preserved in arid or frozen conditions. In the Southwest, paleofeces have been recovered singly from midden deposits or in large quantities from room blocks that were used as latrine areas. In a latrine situation, paleofeces may be distinguishable as separate entities, or they may be found as a large horizon. Excavation of these unique specimens should focus on the recovery of individual samples placed in separately labeled bags. Each sample represents short-time food intake by one individual, so recovery of each individual sample is preferred to the excavation of large clumps of latrine areas that would represent dietary intake of a number of individuals. We can recover DNA and hormones from paleofeces (Sutton et al. 1996; Sobolik et al. 1996; Poinar et al. 2001), providing an excellent database through which to answer research questions on diet, nutrition, and sex determination. For optimal analysis of human DNA

Figure 3.6. Paleofeces from Hinds Cave, southwest Texas. (Courtesy of Stephen Bicknell.)

from paleofeces, the samples must not be touched or otherwise contaminated. Therefore, paleofeces should be collected using sterile latex gloves, and each sample should be placed in a separate, clean, plastic bag to avoid contamination. The samples should not be handled, breathed on, or removed from the bag. Identification and discussion of paleofeces is covered in Fry (1985).

Pollen and phytolith collection should also proceed according to the basic guidelines already defined. Several pollen profile columns should be chosen for sample collection. In addition, other samples can be collected from a variety of features or locations encountered during excavation, such as vessels or grinding implements, to determine what uses the vessels or implements had, and from living surfaces such as room floors to determine room function and potential processing or storage areas.

For example, Bryant and Morris (1986) analyzed functions of ceramic vessels, metates, and manos recovered from Antelope House, a large prehistoric Anasazi Puebloan site located in Canyon de Chelly, through pollen analysis of matrix from the bottom of forty-two vessels and matrix adhering to one metate fragment and five manos. In addition, fourteen soil samples were analyzed for comparison so that natural, environmental pollen could be differentiated from culturally deposited pollen. The authors were able to determine potential uses of several vessels and grinding stones, indicating food preparation or storage of maize, cottonwood, beeweed, cattail, and goosefoot or amaranth. They state that vessels that had been interred in matrix surrounding hearths tended to contain culturally deposited pollen from food preparation or storage, whereas vessels found in midden deposits tended to contain natural, environmental pollen that collected in the samples through postdepositional processes.

Room function can be determined in part through analysis of pollen from matrix samples recovered from room floors. For example, Bryant and Weir (1986) analyzed numerous pollen samples collected above, on, and below floor surfaces of sixteen suspected storage rooms, eleven suspected habitation rooms, and six suspected ceremonial rooms at Antelope House. Most of the storage rooms contained large amounts of economic or culturally deposited pollen such as maize, beeweed, and cattail, indicating possible storage of these plants. Habitation rooms did not contain significant amounts of economic pollen and were thus easy to distinguish from storage rooms. Suspected ceremonial rooms usually contained significant amounts of pine, juniper, and cattail pollen, a distinctly different assemblage than pollen spectra from habitation and storage rooms. The authors state that pollen analysis should be combined with other archaeological information to assess room function.

Collecting pollen and phytolith samples from living surfaces involves a systematic procedure in which samples are taken evenly across the surface, such as every ten centimeters (figure 3.7). In the example provided, I collected pollen samples from the floor of a pithouse from Old Town Ruin, LA1113, New Mexico. As excavation proceeded through the upper fill of the pithouse, which happened to be an overlying Mimbres room block, one-hundred- to two-hundred-milliliter matrix samples were collected from the surface of the pithouse floor. To avoid modern pollen contamination, samples were collected as soon as that area of the floor was exposed, using a newly cleaned spoon or trowel. As excavation proceeded and more of the pithouse floor was exposed, sampling continued until the entire floor was exposed and sampled. Samples were labeled and identified according to a grid system in which one wall (y-axis) of the pithouse was labeled alphabetically, and another wall (x-axis) was labeled numerically (figure 3.7). Pollen sample collection proceeded normally through two hearths encountered during excavation and from which flotation and archaeomagnetic dating samples were taken.

In addition to the flotation and fine-screen and the pollen and phytolith samples collected from a variety of contexts within the Puebloan site, samples were also collected around the site to compare human versus natural deposition.

SHELL MIDDEN TYPE SITE

Type site 2 is a large shell-midden site on the coast of northeastern North America. Periodically through time, people lived at the site and collected large quantities of mollusks. The discarded mollusk shells formed a large shell midden in which other aspects of prehistoric life such as living floors, hearths, and discarded animal bones were preserved. The site contains a huge faunal assemblage in which the best-preserved fauna (mollusk shell remains) are probably overrepresented. The shells, however, produce an alkaline environment, due to the calcium carbonate content of the shell, in which bone remains are well preserved. Carbonized botanical remains are also recovered, although their preservation is not as good as bone.

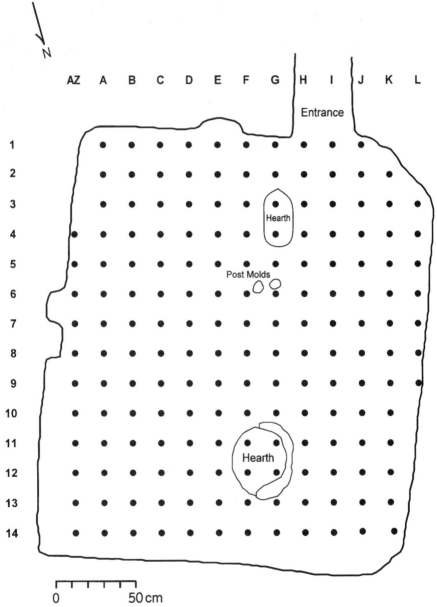

Figure 3.7. Example of pollen sampling strategy from the floor of a pithouse. Each block represents a 100- to 200-ml pollen soil samples. (Courtesy of Stephen Bicknell.)

The stratigraphy of shell middens can be confusing because cultural deposition developed as humans dumped refuse in ever-changing areas of the site (figure 3.8). People shifted living and dumping areas on and around the midden through time. Because of this, the stratigraphy of a shell midden is not uniform across the site. This makes determining cultural zones, areas, or levels difficult, particularly during excavation. Therefore, archaeobiological samples should be collected in a variety of areas in such a site and will be useful for determining cultural areas, areas that are usually determined after complete analysis of all associated artifacts.

In such a confusing site, archaeobiological samples should be collected according to the basic sampling procedure (figure 3.5; table 3.1) in which several 1 × 1–meter units are chosen for systematic two-liter matrix collection from the southwestern quad. Several units are chosen for systematic collection, and as the excavation area expands, more archaeobiological samples are collected. Most areas of a shell midden site, of course, consist of shell with some nonshell matrix. Collecting matrix samples can be difficult, so large pieces of shell should be removed so that a representative two-liter sample for flotation and fine screening can be obtained from all areas of the site chosen for sample collection. This may mean that the southwestern quad of each level (probably representing arbitrary five- or ten-centimeter levels) will be entirely included in the fine-screen and flotation sample after large pieces of whole shell have been removed.

Shell midden sites may contain hearth remains, living floor surfaces, burials, and remains of occupation or storage areas. Archaeobiological samples should be collected from all of these features with particular emphasis on samples in which midden deposits are not mixed with feature deposits. If the feature is large, more than two liters of matrix can be collected as long as the volume is recorded.

Pollen and phytolith samples can be collected from shell midden sites using the basic sampling procedure (table 3.1), although the alkaline depositional environment of a shell midden is not conducive to their preservation. Because of the presence of a large number of shells, collecting pure matrix deposits will be difficult if not impossible. Collection may need to be limited to features that contain little or no shell or areas of the site that contain less shell. Because shell midden sites mainly reflect cultural deposition and modification by humans, pollen and phytolith samples from these sites should not be used to determine paleoenvironment.

Figure 3.8. Stratigraphic profile of the Todd site, a large shell midden on the coast of Maine. (Courtesy of Stephen Bicknell and David Sanger.)

ROCK-SHELTER TYPE SITE

Cultural deposition in rock-shelters is very similar to that observed in shell midden sites. Stratigraphy and depositional patterns can be very confusing due to intense human occupation and refuse disposal and accumulation. Rock-shelters tend to exhibit many noncultural taphonomic factors that affect and modify deposition, making assessments of cultural zones, areas, or levels even more difficult than in shell middens. Because of these difficulties, it is imperative that archaeobiologists are involved in excavation and sample collection so that they can understand and help interpret taphonomic factors influencing the archaeobiological record.

Rock-shelters tend to have excellent preservation of archaeobiological materials, depending on specific environmental conditions, which warrant the time spent on clarifying taphonomic processes. Bone and botanical remains can be abundant. In some cases, such as the dry rock-shelters of the Lower Pecos Region of southwest Texas (Shafer 1986), botanical materials are so plentiful and well preserved as to be comparable to shells in a shell midden (figure 3.9). Uncarbonized perishable remains in particular can be recovered,

Figure 3.9. Archaeobiological remains recovered from the coarse screen of Hinds Cave, southwest Texas. (Courtesy of Kristin D. Sobolik.)

such as basketry, sandals, nets, digging sticks, bone tools, and paleofeces.

Archaeobiological flotation and fine-screening sample collection should proceed according to basic techniques (figure 3.5; table 3.1). Special care should be taken to assess taphonomic factors affecting the assemblage (see chapter 2). Samples should be taken in both potential cultural and noncultural areas so that differences between these two areas can be analyzed and compared. Because rodents introduce noncultural material into the deposit and move cultural material out of context, rodent burrows should be excavated and collected separately. Fine-screen and flotation samples can be collected from rodent burrows but should be considered in their context and should be used to analyze rodent activity patterns rather than cultural lifeways. Rodent activity studies are useful because they help determine site-specific and potentially regional taphonomic factors (see the case study in sidebar 2.1).

Many rock-shelter sites display microstratigraphic depositional patterns that reflect both cultural and noncultural processes. Therefore, archaeobiological samples should be collected from these cultural and natural levels, if visible, rather than from arbitrary levels that may cross-cut these strata. Samples collected from arbitrary ten-centimeter levels could reflect both natural and cultural deposition or cultural deposition from various occupations or activities.

Pollen and phytolith samples can be recovered according to basic sampling procedures (table 3.1). As with flotation and fine-screen samples, pollen and phytolith samples should be taken from a variety of contexts throughout the site, as well as from contexts away from the site for comparison of cultural and noncultural deposits. Pollen and phytolith sampling also needs to follow natural or cultural stratigraphy, if present, which may mean taking more samples than one every ten centimeters, as microstratigraphy warrants.

OPEN TYPE SITE

This type site is an open site located in plains, desert, or wooded areas with evidence of erosion and surficial modification such as plowing, tree throws, digging, or animal burrowing. The site may contain lithic scatters and evidence of hearths (e.g., fire-cracked or burned rocks or stained soils; figure 3.10). Remains of cultural activity at the site tends to consist of durable artifacts and features. Preservation of

Figure 3.10. Fire-cracked rock representing the remains of an eroded, prehistoric hearth at BIBE 57, Big Bend National Park, Texas. (Courtesy of Kristin D. Sobolik.)

organic material tends to be poor because much of the site is exposed and has been eroded or modified by cultural and noncultural agents, degrading organic material. Many sites located through CRM projects fall into this category.

Archaeobiological sampling should take place according to standard procedures (figure 3.5; table 3.1), although much of the site may consist of noncultural matrix surrounding sporadic cultural deposits represented by durable artifacts and features. Particular emphasis should be placed on sample collection in and surrounding features. Samples collected from features may include postdepositional fill but also generally contain remains of prehistoric activity. Collecting samples around features will allow you to compare samples to determine whether the feature fill is mostly cultural or postdepositional. If samples from the feature can be ascribed to cultural deposition, charcoal from these samples is an excellent source for radiocarbon dates.

At the Hulme site, an open Central Plains Tradition (CPT) homestead in central Nebraska, a large surface scatter of animal bones and artifacts was noticed during field leveling (Bozell 1991). Most of the site was excavated using quarter-inch screens in which large bones from antelope, deer, and bison were recovered. Flotation and fine-screened samples were taken from the central fireplace, and four trash-filled storage pits from which bones from a smaller and more varied fauna were recovered, such as snake, prairie dog, turtle, and fish. Analysis of the full set of remains indicates that people had a much more varied meat intake than previously hypothesized. Analysis also suggests that during the CPT, the climate became drier, so the grass forage was less preferable for bison and more preferable for antelope. Plains bison herds may have declined in number due to increasingly arid climate and human hunting patterns.

Pollen and phytolith samples can be collected from open type sites according to standard procedures (table 3.1), and, as with flotation and fine-screen samples, particular emphasis should be placed on cultural deposits such as features. Samples collected from around features and off site also are useful for comparison with cultural deposits as well as for paleoenvironmental reconstruction. Phytolith analysis is particularly useful in grassland environments because grass taxa cannot be distinguished on the basis of pollen analysis, while grass phytoliths are identifiable to various levels.

BURIED TYPE SITE

Buried sites are often hard to locate (see Toolkit, volume 2) but may contain well-preserved archaeobiological materials. Buried deposits are protected against surficial modifications that are prevalent in open sites. Buried sites also tend to have good stratigraphy, which permits defining cultural and noncultural deposits. For example, the Devil's Mouth site is a large alluvially buried terrace site located at the confluence of the Devils River and Rio Grande in Texas (figure 3.11). The site was discovered during archaeological survey for the Amistad Dam, a six-mile-long structure built for hydroelectric power, flood control, and recreational purposes. Impounded waters behind the dam created a 64,860-acre reservoir that extends seventy-three miles up the Rio Grande, fourteen miles up the Pecos River, and twenty-five miles up the Devils River (Labadie 1994). Impounded waters cover at least six hundred archaeological sites, including the Devil's Mouth site. The site is at least thirty-six feet thick, with twenty-four recognizable strata of cultural midden deposits alternating with silt, sand, and clay deposits (Johnson 1964).

Figure 3.11.   Excavation of the Devil's Mouth site, an alluvially deposited site at the juncture of the Devils River and Rio Grande.

Because alluvially buried sites are located along waterways, they can be impacted (destroyed, eroded, or removed) by impoundments for energy sources or recreational facilities. Alluvially deposited sites can contain a large amount of noncultural matrix; therefore, archaeobiological sampling should focus on cultural deposits and features. Analyses of archaeobiological samples from cultural strata and from hearths, pits, and occupation floors will provide information on changing patterns of plant and animal use at the site (see the case study in sidebar 1.1). In addition, analysis of noncultural matrix samples allows comparison of paleoenvironmental conditions before and after cultural deposition. At a site like the Devil's Mouth site, two-liter archaeobiological samples should be taken for flotation, and fine screening from the southwestern quad of a number of 1 × 1–meter excavation units continuously through cultural and noncultural strata, so that changes in environment, as well as cultural patterns, can be ascertained.

Pollen and phytolith samples should be obtained through the standard procedures (table 3.1). It is important to make sure samples from cultural deposits are distinguished from those from noncultural deposits. Analysis of pollen column samples that include cultural and noncultural deposits will provide a good framework for tracing changes in environment through time as well as human impacts and plant use. Pollen analysis of matrix samples collected in column fashion from all twenty-four strata from the Devil's Mouth site revealed several climatic changes in the region over time (Bryant 1966). From 7000 to 3000 B.P., the region had a mesic environment that was probably caused by increased rainfall during spring and summer. Erosion of Rio Grande terraces occurred at the end of this mesic interval (3000–2500 B.P.), which was followed by an increasingly xeric environment up to the present.

CONCLUSION

Excavation and recovery of archaeobiological samples, including pollen and phytolith samples, can follow a standard procedure. This procedure can be modified on the basis of site type, environmental conditions, time and money constraints, and research design. It is a good idea to collect more samples than needed and even more samples than will actually get analyzed with your particular project. Although curation space is limited, excavation is destructive, and you usually don't get a second chance. In fifty years, someone may need those samples to help answer a research question, or particular samples from Phase II may be needed during Phase III. Having matrix samples sitting on a shelf for possible future analysis is preferable to having samples lost in the backdirt pile.

4

LABORATORY AND
ANALYTICAL TECHNIQUES

Archaeobiological analyses require a significant amount of technical expertise that is learned through training, experience, and
many, many hours of analysis. Archaeobiological analyses should only
be conducted by trained archaeobiologists (table 1.1). Because these
studies are time-consuming and demand a lot of experience, many archaeobiologists become specialists or experts in particular areas or on
particular botanical or faunal taxa. Technical expertise is the backbone
of archaeobiological analyses, and technical identification and analysis can be the most time-consuming and often tedious step. Anyone
can put in the hard work necessary to become a technical expert in archaeobiology, but to become an archaeobiological scientist involves
using botanical or faunal remains to answer broader-ranging questions.
The most rewarding aspect of archaeobiology is not when a certain
fish bone has finally been identified (although such small victories are
exciting) but when the identification of an entire fish assemblage leads
to new discoveries, answers previously unanswered questions, or indicates that a modification of a hypothesis is necessary.

For example, McInnis (1999) chose prehistoric subsistence practices in coastal Peru for her master's thesis topic. She helped conduct excavation at Quebrada Jaguay and chose to analyze the faunal
remains. Overall, the remains were poorly preserved and consisted
mainly of fish fragments and a large number of otoliths (fish ear
bones). Identification of the fish bone fragments would have taken
an immense amount of time and energy, from which McInnis probably would only have been able to identify one or two species or

$$Y = aX^b \quad \text{or } \log_{10} Y = b(\log_{10}X) + (\log_{10}a)$$
$$Y = \text{Standard Length}$$
$$X = \text{Otolith Length}$$
$$a = Y\text{-Intercept}$$
$$b = \text{Slope}$$

Figure 4.1.   Allometric scaling formula. (From Reitz and Wing 1999: 175–76.)

fish elements. Instead, she focused on the otoliths and conducted an analysis of the size and type of fish taken at the site using regression formulas provided by Reitz and Wing (1999) (figure 4.1). Using the regression formulas and size of the otoliths, McInnis estimated population structure of the drum fish (Sciaenidae) at the site (figure 4.2).

The technical portions of her analysis complete, McInnis, in conjunction with the project director, used her data to address issues regarding early maritime adaptations, revealing that Paleoindian-age people in South America were adapted to marine-based environments and that Quebrada Jaguay is the earliest known site providing such evidence (Sandweiss et al. 1998). McInnis's work reveals the importance of technical analyses that ultimately feed into larger issues and broader-ranging research questions. The first steps toward being able to answer such questions, however, are technical steps.

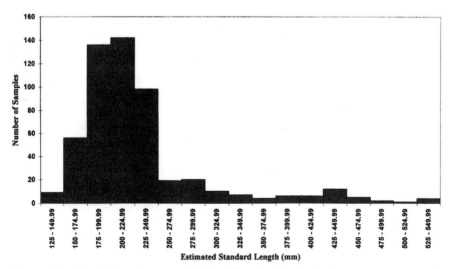

Figure 4.2.   Population structure of Sciaenidae for Quebrada Jaguay. (From McInnis 1999.)

REFERENCE COLLECTIONS

The first and most important step in archaeobiological analysis is generation of an extensive and usable reference collection of modern plants, animal bones, and shells for comparative purposes. Archaeobiologists should be familiar with the plants or animals in a region and should have an extensive reference collection from that area before proper identification and analysis of archaeological samples can take place. One reason archaeobiologists should be involved in overall project design and be present during excavations is so they can collect modern reference samples from the surrounding area.

For paleoethnobotanists, developing a collection involves collecting a wide variety of whole plants during different seasons so that the life cycle of the plant is represented. Whole plants can be collected in a plant press and dried for preservation. Be sure to record where and when each plant was collected. Other reference samples should include pieces of wood (some of which can be charred for charcoal identification), nuts, seeds, berries, roots, and phytoliths from all portions of each plant. Pollen reference samples are also collected from individual plant flowers during the appropriate season. Chemical processing of modern plants needs to be undertaken to remove pollen (steps IV and V of table 4.2) and phytoliths (steps III and IV of table 4.3) for reference samples. The most important part of collecting modern plants to use as comparative reference samples is to definitively identify them. Otherwise, they are useless or potentially problematic. Be sure to consult a qualified botanist if there are any questions regarding plant identification.

For zooarchaeologists, animals can be obtained, usually with an appropriate permit, via roadkill collection, trapping, and taxidermy businesses. Animals on endangered or threatened species lists often cannot be collected, even through roadkill. A list of these animals, as well as collection permits, usually can be obtained through the state departments of natural resources or similar state agencies. Keep in mind that reference sample collecting requires defleshing animals to recover all bones or shells. This process can become an interesting exercise in ethnoarchaeology, or at least a learning experience in anatomy and morphology. Be sure to work with someone experienced in butchery and defleshing before cutting or macerating on your own. Also, record information on the animal's size, weight, sex, age (if possible), and other life characteristics.

Reference collections in museums and laboratories can supplement your collection. For example, I use the faunal collection curated at the

University of Maine Zooarchaeology Laboratory in addition to my own personal zooarchaeological reference collection. To expand my reference sample for species not available in either of those collections, I use collections at the Museum of Comparative Zoology at Harvard University. Analysts should know where such collections are located and use them if necessary.

In addition to comparative reference collections, a number of identification guides and atlases are available to help identify both plants and animals. Guides and atlases are useful for identification and should be used in conjunction with, but never in place of, a comparative reference collection. I cite identification guides in each appropriate section.

After an extensive reference collection has been accumulated or located, archaeobiological identifications can proceed using those collections in addition to the age-old process of lots and lots of time and experience. I teach a zooarchaeology class in which each student learns basic faunal identification skills, identifies a small faunal collection from an archaeological site, writes a faunal report, and presents the information to the class. Although I can lead the students in an appropriate direction or point out easy identification tips, the students are usually surprised to realize that learning identification is mainly up to them. They need to spend an enormous amount of time familiarizing themselves with animal bone and matching their archaeological bone fragments to bones from the comparative collection. It is a time-consuming and often tedious process in which there are no shortcuts. At the end of the class after their final paper is turned in, many students feel that they are now zooarchaeologists. Yet this step is only the very first stage in becoming technically competent. With further experience comes more expertise. Most archaeobiologists have probably had to reanalyze their first few research projects, correcting a number of technical errors along the way. If they haven't done this, they probably should; it is an enlightening experience. Archaeologists contracting for or hiring archaeobiologists should realize that the best analysts are usually the most experienced—and expensive—analysts.

BOTANICAL REMAINS

Botanical remains that are recovered during coarse screening should be sorted into similar categories or groups, such as seeds, charcoal, fiber, and nutshell, from which identification can proceed. This initial sort can be conducted by someone other than the archaeobiologist. All of the material should retain the particular field sack or lot number assigned in the field. Samples should not be washed or modified through brushing or removing adhering matrix unless the analyst deems this necessary.

The most tedious and time-consuming aspect of botanical analysis usually is sorting flotation samples. To facilitate sorting and quantification, samples can be graded into different size classes by screening the samples through nested geological sieves (e.g., two and one millimeter). Samples should be sorted under a magnifying lens or dissecting microscope (10×–20×). All botanical remains should be sorted into groups such as seeds, wood charcoal, fiber, leaves, and miscellaneous unidentified items. They can then be identified as to taxon and element using comparative samples and identification guides. Some useful identification guides for botanical remains include Appleyard and Wildman (1970), Catling and Grayson (1982), Core et al. (1979), Corner (1976), Dimbleby (1978), Gunn et al. (1976), Leney and Casteel (1975), Martin and Barkley (1961), Montgomery (1977), and Western (1970). See Pearsall (1989:182) for references on identification of domesticates. The samples can then be quantified to aid in analysis and interpretation. Several paleoethnobotanical recording techniques are provided in Pearsall (1989:108–22; 2000).

QUANTIFICATION TECHNIQUES

Quantification of plant remains is a significant aspect of archaeobotany. Researchers apply many different quantification methods, and few comparative papers have been published on the quantification procedures and their assessment (but see Hastorf and Popper 1988). Several of the most commonly applied procedures are discussed here.

The presence/absence or ubiquity method simply measures how many samples contain each taxon within a group of samples. Either a taxon is present in a sample, or it is absent. No matter what other quantification methods are employed, every study should use presence/absence. Because it is a universally used technique, this form of quantification permits easy comparison between samples and assemblages. Presence/absence reduces the effects of differential preservation and sampling, although the number of samples and taxa within a sample will affect results: The more taxa that are recognized in a sample, the more important a common botanical constituent will seem and the less important an infrequent botanical constituent will seem. This is also the case as the number of samples is increased.

In the *percentage weight method,* all of the botanical constituents in a sample, including both flotation and coarse-screen samples, are separated and weighed. The weights are compared directly or are reflected as a percentage weight of the total. The weight technique is often used in quantification, thus making analyses using this method easier to compare with other studies. The major drawback of this method is that it underestimates the lighter contents, such as fiber, and overestimates the heavier contents, such as wood charcoal.

Another frequently used quantification technique is the *percentage count method,* in which the botanical remains are counted and compared to the total botanical count. This method tends to overestimate botanical remains that are easily broken or contain more fragments to begin with, such as fiber particles and wood charcoal. However, this method is not time-consuming, relatively easy to conduct, and additive, so that when new samples have been quantified the percentage count simply can be added to the total (similar to the NISP [number of identified specimens] quantification method for faunal remains).

In the *percentage volume method,* all material from the sample is separated and placed into standard size containers. The number of containers each constituent fills is then compared to the total. This technique is fairly sufficient in estimating the amount of each item in a sample, although this method is very cumbersome and inexact. It is almost impossible to get irregularly shaped botanical remains, such as strands of fiber, nutshells, and wood charcoal, to fill all volumetric space of standard size containers. This method, thus, uses more "guestimation" than other quantification methods employed. It is also not widely used in archaeology, making comparisons difficult.

For the *percentage subjective method,* all of the botanical constituents in a sample are aligned according to their frequency, from most frequent to least frequent. These frequency groupings are then placed into percentage groups that provide a range of error. Each constituent is placed into these different percentage groups when the sample is being sorted and separated, making it the least time-consuming and most cost-efficient quantification technique. The percentage subjective method does not overestimate larger items or items that are broken into numerous pieces. The problems with this technique are threefold: (1) Quantities are presented as a range so data cannot be manipulated statistically; (2) this technique is not additive so that when more samples are added to the original analysis, quantification percentages cannot be added to the previous total; and (3) this technique is not widely used, making comparison between different studies often difficult.

I used the percentage subjective method in an analysis of thirty-eight paleofecal samples from Baker Cave in southwest Texas (Sobolik 1988). This quantification method was chosen because it is the easiest and introduces the least amount of bias into the analysis. The percentage subjective method does not overestimate or underestimate the importance of any botanical remains. The technique is also useful because each paleofecal sample represents an individual analysis unit; therefore, the problem of adding samples to a total was not encountered. A secondary quantification technique, the weight method, was also used for comparative purposes and to facilitate any future statistical analyses. The weights were recorded but were not analyzed because the weight method overestimates the importance of heavier botanical items and understimates lighter botanical items. The botanical remains from each paleofecal sample were sorted, identified, and visually assessed to determine the relative percentage of each constituent as compared to the total sample (table 4.1). For example, I estimated that sample 5 contained 80 to 94 percent prickly pear fiber; 1.4 percent onion fiber, yucca fiber, sotol fiber, and prickly pear seeds each; and a trace amount of mustard seeds. The percentage subjective method provided a quick, efficient method of quantification that did not overestimate or underestimate any botanical constituent, and it could be done at the same time as the sample was sorted.

POLLEN ANALYSIS

Pollen analyses from archaeological sites and other environmental conditions can offer a diversity of information on prehistoric populations and subsistence practices that cannot be determined through the sole analysis of other archaeobiological remains. Pollen is observed and analyzed from many sample types. Such diverse uses for pollen include paleoenvironmental reconstruction, archaeological dating techniques, and paleodiet. Pollen is useful in a wide range of areas, mainly due to its prevalence in the environment and its distinctive, sturdy structure.

Pollen is a sturdy structure due to its exine (outer layer), which is composed partly of sporopollenin, a strong, resistant substance. The inner layer of pollen (intine) consists of cellulose, which is

Table 4.1. Percentage Subjective Quantification of Some Botanical Remains from Eight Paleofeces from Baker Cave, Texas

Botanical Remains	Sample Number							
	1	2	3	4	5	6	7	8
Fiber								
Onion		D	F	H	H	E	G	H
Prickly pear	B		H	G	B	H		H
Yucca					H			
Agave						H		H
Sotol			H	I	H			
Seeds								
Prickly pear					H		A	B
Juniper							H	H
Cactus						H		
Mustard					I			
Nutshell								
Acorn	H							
Walnut				G				

A = 95%–100%
B = 80%–94%
C = 65%–79%
D = 50%–64%
E = 35%–49%
F = 20%–34%
G = 5%–19%
H = 1%–4%
I = 0.1%–0.9% (trace)
Source: Modified from Sobolik (1988).

easily degraded after a short length of time, such as in archaeological deposits. When the cellulose layer of the intine is degraded, only the outer layer containing sporopollenin remains. However, this layer is often sufficient for identification because the exine contains distinct sculpturing patterns and aperture shapes (figure 4.3), allowing for pollen identification to be made to the species level in some instances.

Pollen types are divided into insect-pollinated plants (zoophilous) and wind-pollinated plants (anemophilous). Insect-pollinated plants produce relatively few pollen grains and are usually insect-specific. Insect-pollinated plants generally produce fewer than ten thousand pollen grains per anther (Faegri and Iversen 1964). These pollen types are rarely observed in the pollen record due to their low occurrence in nature and method of transport.

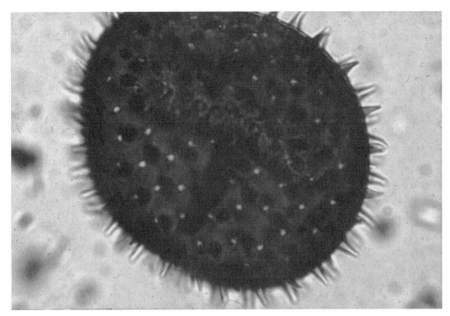

Figure 4.3.   Malvaceae pollen grain showing the echinate (spiny) patterning of the exine. (Courtesy of Kristin D. Sobolik.)

Wind-pollinated plants, on the other hand, produce large amounts of pollen and are frequently found in the pollen record. Faegri and Iversen (1964) state that an average pine can produce approximately 350 million pollen grains annually, and Mack and Bryant (1974) found pine pollen percentages over 50 percent in modern deposits where the nearest pine tree was more than a hundred miles away.

A high frequency of wind-pollinated pollen types in archaeological samples most likely indicates natural environmental pollen rain rather than human subsistence activities. High frequencies of insect-pollinated pollen types, however, may indicate human use and modification of that particular plant.

Pollen analysts process samples in various ways, although the basic procedure involves removing organics, silicates, and carbonates. A standard procedure to extract pollen from soils is provided here (table 4.2). I do not recommend that untrained technicians conduct this procedure. Sample preparation involves use of highly dangerous chemicals and requires specialized equipment and safety protocols. Pollen extraction should be done only by trained technicians who realize the potential dangers at each step and take appropriate precautions to avoid damage to person, property, and pollen.

Table 4.2.   Standard Pollen Extraction Procedure*

Step I: Removal of large organic or mineral particles
1. Remove thirty to fifty milliliters of soil from the sample collected. If samples come from heavily weathered areas or alluvial sediment, use one hundred milliliters of soil.
2. Screen samples through a one-millimeter mesh screen into a beaker. Discard material caught in screen.
3. Add one to two *Lycopodium* spp. spore tablets, carefully recording number of spores per tablet.

Step II: Removal of carbonates
4. Add concentrated HCl (38 percent) to remove carbonates and dissolve calcium bonding in spore tablets. Stir and allow reaction to take place. If reaction causes foam, use fine spray of ethanol to disperse.
5. Pour off and discard liquid fraction. Add a thousand milliliters of distilled water to sediment in beaker and stir. Let solution settle for two hours. Repeat this step two more times. Place remaining sample in fifty-milliliter centrifuge tubes.
6. Centrifuge the residue at 2,000 RPM for fifteen seconds. Discard liquid fraction.

Step III: Removal of silicates
7. Transfer remaining sediment into plastic beakers and add small amounts of 70 percent HF** acid until matrix sample is covered. Stir occasionally and let sit overnight.
8. Add distilled water to beaker and stir. Let solution settle for two hours. Pour off and discard liquid fraction in fume hood sink. Repeat this step at least two more times. Place remaining sample in fifty-milliliter centrifuge tubes.

Step IV: Removal of organics
9. Rinse residue in glacial acetic acid to remove water. Centrifuge and decant.
10. Prepare acetolysis mixture: nine parts acetic anhydride, one part sulfuric acid.
11. Add acetolysis mixture to samples, stir thoroughly, and place in a boiling water bath for five minutes. *Don't mix water from water bath with acetolysis mixture!* Remove, centrifuge, and decant. Repeat.
12. Wash sample in distilled water. Centrifuge and decant. Repeat.

Step V: Slide preparation
13. Place remaining residue in a small vial with glycerine for curation. Label.
14. Take a small portion of glycerine mixed residue and place on a microscope slide. Place coverslip over sample and secure with nail polish or other sealant. Identify and count pollen.

*Pollen extraction techniques involve the use of toxic chemicals. Extraction should never be attempted without a fully functioning fume hood and protective coat, gloves, and goggles. Processing should be done by a trained technician.
**Use of HF must be restricted to a fume hood. HF fumes are very harmful and can cause permanent damage to lungs and nose if inhaled. Contact with HF can be fatal so always wear plastic coat, plastic gloves, and plastic face mask.

As with other identifications, pollen identification must proceed using a modern comparative collection of representative pollen types from the region. Learning pollen identification is time-consuming and involves practice and experience. A standard North American identification key is Kapp (1969) and Jones et al. (1995).

Pollen data can be presented in various ways, but most data are presented as a pollen diagram. An example of a very basic pollen diagram is presented here (figure 4.4). In this diagram, cultural zones or stratigraphic levels are listed on the left axis, and associated radiocarbon dates are provided on the right axis. Most diagrams present data in stratigraphic and chronological order with the bottom of the diagram representing the deepest (and hopefully oldest) deposits and the top representing the youngest and, in many cases, modern deposits. Pollen taxa are listed along the top border with their observed percentages in each sample provided in black. Included in this particular diagram is a depiction of the AP (arboreal pollen) versus NAP (nonarboreal pollen). Because this sample represents a stratigraphic profile or column, the data are presented as a change through time, and the pollen percentages are filled in black from one sample (or pollen zone) to another.

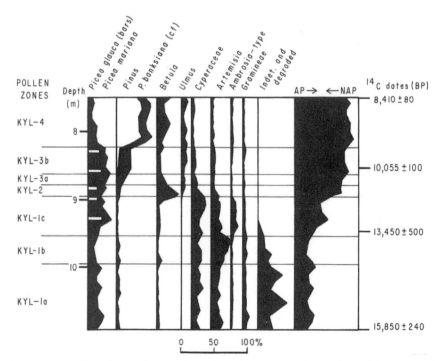

Figure 4.4.   Pollen diagram from Kylen Lake, Minnesota. (Redrawn from Birks 1981 and taken from Holloway and Bryant 1985)

In another example (figure 4.5), the pollen diagram illustrates pollen identified from individual paleofecal samples rather than from a continuous profile. This type of diagram also can be used when presenting individual pollen samples, such as surface samples or samples from archaeological features, rather than stratigraphic column samples. In this diagram, the individual samples and associated radiocarbon dates are presented on the left axis. Pollen taxa are presented across the top and their percentages as observed in each sample are illustrated. In this particular case, crop pollen is separated from wild plant pollen and pollen concentration values are listed.

The amount of pollen present in a particular volume of soil can be estimated using pollen concentration values. Pollen concentration values are determined through the ratio of pollen grains to *Lycopodium* tracer spores added to each sample before processing. The number of spore grains added is multiplied by the number of prehistoric grains counted in the sample. This number is divided by the number of spore grains counted multiplied by the amount of sediment processed. Concentration values help determine the amount of prehistoric pollen present in a sample and can help assess depositional rates for soil samples and possible pollen ingestion for paleofecal samples.

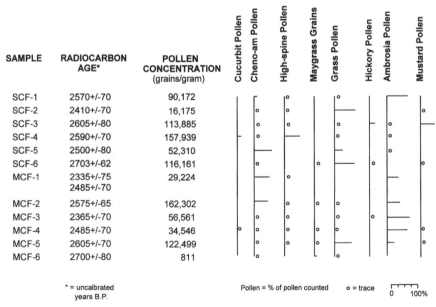

Figure 4.5.   Pollen diagram from Mammoth Cave paleofeces. (Modified from Gremillion and Sobolik 1996.)

PHYTOLITH ANALYSIS

Phytoliths, meaning plant stones, are "particles of hydrated silica formed in the cells of living plants that are liberated from the cells upon death and decay of the plants" (Piperno 1988:1). They are formed in plants after siliceous or calcareous substances from groundwater are absorbed and deposited in numerous cells throughout the plant (figure 4.6). Phytoliths composed of calcium oxalate also form in certain cacti and desert succulents (Jones and Bryant 1992; figure 4.7). When the plant dies and decays, these durable calcified or silicified cells remain in the soil as indicators of the plant in which they were formed. As with pollen, different plants produce diverse, and often unique, morphological phytolith types. Unlike pollen, different parts of the plant produce morphologically different phytolith types, making the use of comparative collections from all plant parts essential. However, in some areas in which pollen types are not distinctive (such as with grasses), phytolith analyses can produce excellent results. Phytoliths also can be preserved in environmental conditions in which pollen is degraded or absent, and can be used in conjunction with pollen analyses for a more complete paleoenvironmental and

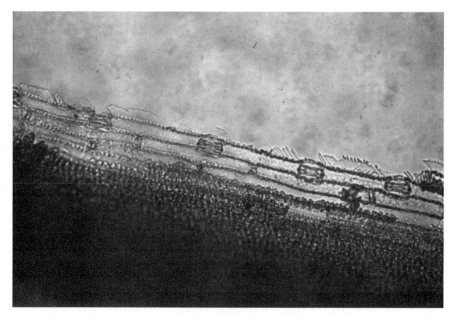

Figure 4.6.   Grass fiber fragment showing dumbbell-shaped phytoliths. (Courtesy of Kristin D. Sobolik.)

Figure 4.7.   Calcium-oxylate druse crystals, a form of phytoliths, from prickly pear cactus.
(Courtesy of Kristin D. Sobolik.)

archaeological picture. For information on identification and interpretation of phytoliths, see Piperno (1988), Piperno and Pearsall
(1993), and Rapp and Mulholland (1992).

An example of such an integrative pollen/phytolith study was conducted by Cooke and colleagues (1996) in an attempt to reconstruct
the prehistoric Panamanian landscape. In this integrative analysis,
vegetation history (as determined through pollen, phytoliths, and
charcoal), archaeological faunal remains, and isotopic analyses of human bone were combined to provide an overall picture of changing
human subsistence patterns and environmental modifications in
Panama. Vegetation history was determined through the analysis of
sediment cores taken from three lakes and swamps. Pollen and phytolith analyses were combined with wood charcoal research to indicate environmental changes from vegetation adapted to a cooler, dry
environment during Paleoindian occupation, to extensive environmental modification when slash-and-burn horticulture became prevalent. Phytolith data also indicate that palms and arrowroots may have
been cultivated in hill slope gardens and that maize started to become

prevalent about 5000 to 3000 B.P. Faunal analyses and isotopic composition of human bone indicate that diet differed among groups that occupied different sites and environments.

Phytolith analysts process samples in various ways. Side-by-side tests using various processing techniques can help determine which works best under which conditions (Pearsall 1989). All phytolith processing techniques involve floating phytoliths from matrix using heavy density separation. While heavy density separation steps vary among analysts, the standard procedure used by Piperno (1988) is listed in table 4.3. Phytolith processing and analysis should only be conducted by trained archaeobiologists.

ZOOARCHAEOLOGY

Faunal remains are recovered from archaeological sites through coarse and fine screening. Bones recovered during coarse screening should be sorted as a separate category from other artifacts and placed into a separately labeled bag. All of the material should retain the particular field sack or lot number assigned in the field. It is not recommended that untrained archaeologists or field technicians presort bone into other categories, such as fish, bird, and mammal. In my experience, such presorts end up taking more time and energy and being redone later as it is determined that many of the mammal bones are in fact bird bones and vice versa. Trained archaeobiologists can easily determine which bone is mammal, bird, or fish, therefore eliminating wasted time and resources in presorts.

Samples should not be washed or modified through brushing or removing adhering matrix unless the analyst deems this necessary or in cases in which matrix is adhering to the bone, hindering identification and observation of surficial modifications. Time and energy is wasted washing every bone recovered from an archaeological excavation. Not only does this step waste time and energy, but it may cause bone to break or crumble and may add marks on the bone that can obscure prehistoric modifications such as cut marks and gnaw marks. In fact, Sutton (1994) recommends that most artifacts, including archaeobiological remains, should *not* be washed because we may be washing away important evidence such as organic and protein residues. And with our increasing awareness of nonvisual remains associated with archaeological samples, such as DNA, it becomes clear that the less we handle and modify any archaeological material, the better.

Table 4.3.   Standard Phytolith Extraction Procedure*

Step I: Separation of phytoliths and removal of clay
1. Defloculate soil samples with 5 percent solution of Calgon or sodium bicarbonate.
2. Screen with fifty-three micron mesh screen. Keep the sample caught in the screen.
3. Place remaining sample (which passed through the screen) in large beakers and add water to three-quarters full. Stir vigorously. Let solution settle for one hour. Pour off and discard liquid fraction. Repeat at least five times.
4. Place remaining sample in hundred-milliliter beakers and add water. Stir, let settle for three minutes, and pour off supernatant liquid into a thousand-milliliter beaker (this separates fine and coarse silt fractions). Repeat, allow to settle for two minutes, pour off supernatant liquid into the same thousand-milliliter beaker. Repeat at least five times.

Step II: Removal of carbonates
5. Place 1 to 1.5 grams of each silt sample and the screened sand sample (three samples total) in test tubes and rinse with distilled water.
6. Add HCl (10 percent) to remove carbonates, centrifuge at 500 rpm for three minutes, and pour off liquid fraction. Repeat until no reaction is observed. Rinse with distilled water.

Step III: Removal of organics
7. Add hydrogen peroxide (3 percent) or concentrated nitric acid to remaining sample. Place in boiling water bath until reaction stops. Repeat.
8. Conduct heavy density separation with zinc bromide, specific gravity 2.3. Mix ten milliliters of heavy density solution into samples and centrifuge at 1,000 rpm for five minutes. Remove liquid (containing phytoliths) to a second centrifuge tube. Remix initial sample and repeat centrifugation. Remove liquid to second centrifuge tube. Repeat if necessary.
9. Add distilled water to liquid portion in 2.5 to 1 ratio. Centrifuge at 2,500 rpm for ten minutes, decant, and discard liquid. Repeat twice.
10. Wash in acetone.

Step IV: Slide preparation
11. Place remaining residue in a small vial with permount for curation. Label.
12. Take a small portion of permount-mixed residue and place on a microscope slide. Place coverslip over sample, and secure with nail polish or other sealant. Identify and count phytoliths.

*Phytolith extraction techniques involve the use of toxic chemicals. Extraction should never be attempted without a fully functioning fume hood and protective coat, gloves, and goggles. Processing should be done by a trained technician.
Source: Modified from Piperno (1988).

Archaeobiologists vary widely on the ways in which they initially process, sort, label, and analyze bone. I have had many chances to observe this diversity in a faunal working group of six zooarchaeologists from Maine, of which I am a member. One of the research projects the group tackled was an analysis of faunal remains from a shell midden site on the coast of Maine (Spiess et al. 1998). Each of us has our particular expertise, so we divided the bone analysis accordingly. One member analyzed the bone tools, one the fish, another the shell chondrophores for seasonality, another the small mammals, and so forth. We spent many hours discussing and arguing about the basic methods we should employ to analyze the bone. What an enlightening experience! Some of the groups wanted to take bone from a separately labeled and recognized cultural unit, sort the bone into different categories, identify taxa and surficial modifications, record these observations, and place all the unlabeled bone back into the initial bag. They indicated that it was too time-consuming and inefficient to label each bone and separate each identified faunal category into a separate bag. Some of the group wanted to take bone from an individually labeled bag; analyze each bone separately; identify taxon, portion of element, age, sex, pathology, and surficial modifications; and record that information on a separate card for each bone (figure 4.8). Each bone would be labeled accordingly and placed in a separate plastic bag with its own card, modifying the original field sack or log number to reflect what bone it was. For example, the first mammal bone from a particular field sack was separately labeled (on the bone and on the bone card) with the original field sack or log number and the addition of M.1. The third reptile bone from that same field sack would have the addition of R.3 to the original field sack or log number, again recording that number on the bone and on the bone card. They indicated that this method was best for bone curation and potential reanalysis. Only in this way could reanalysis be conducted easily and efficiently, since it would be easy to find a particular bone; you wouldn't have to search through a bag full of bone to fine the exact bone to reanalyze. Some of us employed analytical methods somewhere in between these two examples.

It is up to the archaeologist and archaeobiologist to ascertain which method is most efficient and useful for the project goals. In most CRM projects, time and money preservation are essential; therefore, it will be necessary to focus more time on bone identification, analysis, and interpretation, and less on bone labeling and intricate recording. In some research-oriented projects, time and money constraints may not be a problem—for example, in academic projects in which student labor is either cheap or free, and there are no specific deadlines in which the project needs to be completed. In such instances, it may be feasible to spend more time on bone labeling, recording, and curation.

It is not necessary to label each bone from a project, and definitely not the small fragments, as long as the bone stays in the labeled

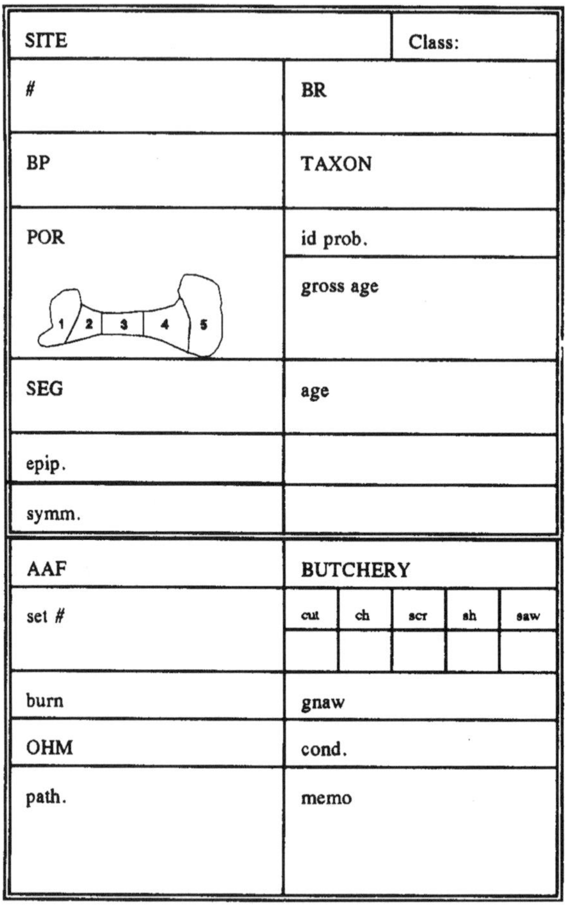

SITE		Class:
#	BR	
BP	TAXON	
POR	id prob.	
	gross age	
SEG	age	
epip.		
symm.		
AAF	BUTCHERY	
set #	cut ch scr sh saw	
burn	gnaw	
OHM	cond.	
path.	memo	

Figure 4.8.   Bone coding card. (Courtesy of Dinah Crader.)

plastic bag that contains the site number and assigned field sack or log number. Some archaeobiologists feel, however, that labeling every bone from a project is essential and should be conducted at the same time that artifacts are labeled. Plastic bags are better to use than paper bags because they will preserve better through time. However, you need to be careful that provenience records written on plastic bags do not wipe off. Permanent markers are necessary for such recording.

Sorting fine-screen samples for smaller faunal remains is almost as time-consuming and potentially tedious as sorting flotation samples. If there is a lot of matrix associated with the fine-screen sample (which has been recovered with flotation samples and essentially water screened), then you may want to pass the sample through nested geological sieves to separate the sample into size grades for ease in sorting and identification. The samples recovered from fine screens are smaller and thus may be harder to identify. A good comparative collection of all size animal species is particularly important for this stage of analysis, including smaller fish, rodents, shrews, bats, reptiles, amphibians, and small birds. The species diversity recovered from a site usually increases dramatically once fine screening has been done (Reitz and Wing 1999). After the remains have been sorted, they can be identified as to particular taxon and element using comparative samples and identification guides. Some useful faunal identification guides include Cannon (1987), Casteel (1976), Claasen (1998), B. Gilbert (1990), B. Gilbert et al. (1985), Olsen (1968), and Sobolik and Steele (1996).

QUANTIFICATION

Studies of quantification in zooarchaeology have been conducted more frequently than in paleoethnobotanical studies, and papers on the subject are more prevalent (Bokonyi 1970; Casteel 1977, 1978; A. Gilbert and Singer 1982; Krantz 1968; Lyman 1979). The most frequently used techniques are ubiquity, NISP, and MNI (minimum number of individuals), all discussed later. Other quantification techniques that have been used include MNE (minimum number of elements), meat weight, and various taxonomic diversity and richness indices (Reitz and Wing 1999). For increased discussion and analysis of these techniques, see Grayson (1984) and Klein and Cruze-Uribe (1984).

The presence/absence (ubiquity) method is inherent in all faunal analyses and allows different samples to be easily compared. Presence/absence reduces the possibility of error of interpretation due to differential preservation of the sample. For example, dense bone such as beaver bone preserves well in most environments, whereas less dense bone such as small bird bone does not. If one small fragment of bird bone and 150 beaver bones are recovered from an archaeological site, the ubiquity index will not distinguish between these two samples. Both are present. The ubiquity method also does not increase the number of sample divisions. As the sample is divided into smaller groups or as sample size is increased, constituents that are more frequent will seem to be more important, whereas constituents that are less frequent will occur in fewer samples and will be considered of minimal importance. Presence/absence information has been proven useful for zoogeography and paleoenvironmental reconstruction as well as dietary purposes. I provide an example of a ubiquity table of the mammal remains from a rock-shelter site in Big Bend National Park, Texas (table 4.4).

Number of identified specimens (NISP), also common to all faunal analyses, is a basic count of the number of bones identified for each taxon category. NISP assessments can be made for an entire site or for any intrasite constituents such as cultural zones or levels, or a combination of different excavated areas; however, total NISP will always be the same for each taxon or faunal grouping used. NISP is additive: When more faunal remains are identified and quantified, the NISP is increased. One drawback to NISP is that it will tend to overestimate the frequency of taxa in an assemblage because it can increase with bone breakage thus inflating NISP. In addition, some animals contain more elements than others, such as turtles and alligators (teeth), and their NISP will therefore be higher. To illustrate NISP, I provide an example of a quantification table of the mammal remains from a rock-shelter site in Big Bend National Park, Texas (table 4.5), the same

---

Table 4.4.   Ubiquity of Mammal Remains from a Rock-shelter in Big Bend National Park, Texas

Artiodactyla
Antilocapra (pronghorn)
Odocoileus (deer)
Carnivora
Mephitis (skunk)
Mustella (weasel)
Vulpes (fox)
Lagomorpha
Lepus (jackrabbit)
Sylvilagus (rabbit)
Rodentia
Cynomys (prairie dog)
Erethizon (porcupine)
Geomys (pocket gopher)
Neotoma (packrat)
Ondatra (muskrat)
Peromyscus (mouse)
Reithrodontomys (harvest mouse)
Sciurus (squirrel)
Sigmodon (cotton rat)
Spermophilus (ground squirrel)

remains on which I previously determined ubiquity (table 4.4).

Another frequently used quantification measure is MNI. MNI measures the minimum number of animals that are represented in a sample by determining the most abundant element of each taxon identified. MNI also may be calculated according to different sides (right and left) of the most abundant element, or by matching elements from individuals, or by sex and age differences. This type of quantification eliminates the possibility of overestimating the number of individuals, which easily can occur when one assumes that each element or fragment represents a different animal. The MNI quantification method is not biased toward animals with many bony parts (e.g., crocodiles, turtles, armadillos), bones that are highly fragmented (e.g., from bone marrow processing), or that were brought whole to the site rather than incomplete (Klein and Cruz-Uribe 1984). I provide an example of MNI from the same mammal remains I used to determine ubiquity and NISP (table 4.5).

Table 4.5.   Quantification of Mammal Remains from a Rock-shelter in Big Bend National Park, Texas

Taxon	NISP	MNI
Artiodactyla		
Antilocapra (pronghorn)	1	1
Odocoileus (deer)	10	1
Carnivora		
Mephitis (skunk)	2	1
Mustella (weasel)	1	1
Vulpes (fox)	4	1
Lagomorpha		
Lepus (jackrabbit)	7	1
Sylvilagus (rabbit)	3	1
Rodentia		
Cynomys (prairie dog)	1	1
Erethizon (porcupine)	4	1
Geomys (pocket gopher)	2	1
Neotoma (packrat)	42	13
Ondatra (muskrat)	1	1
Peromyscus (mouse)	1	1
Reithrodontomys (harvest mouse)	2	1
Sciurus (squirrel)	6	1
Sigmodon (cotton rat)	6	3
Spermophilus (ground squirrel)	2	1
Unidentified mammal	222	
Total	317	31

Several problems arise from the use of MNI. One is that different aggregation techniques will produce different MNI counts (Grayson 1984). As the faunal sample is divided into smaller aggregates, such as for different analytical units or zones, the MNI for each taxon increases because the most abundant element of each taxon could be different for each aggregate. For example, using the rodent remains from the rock-shelter in Big Bend, MNI changes when cultural zones and provenience data are taken into consideration. Previous tables of rodent remains (tables 4.4 and 4.5) provided quantification of fauna from the site as a whole. When rodents are differentiated according to unit designations, rodent MNI changes (from twenty-four to thirty-two rodents), although ubiquity and NISP (sixty-seven) remain the same (table 4.6).

Another problem with the MNI method is that animals that occur in low numbers will tend to be overestimated while more commonly represented animals will be underestimated. When one bird bone is observed, the MNI for bird is one. If ten different rabbit bones were observed the MNI for rabbit would also be one even though there is a high probability that the rabbit bones are from more than one animal. Therefore, animals that occur frequently will be underestimated in the MNI tabulation.

Another problem with the MNI method is that different investigators determine the MNI differently. Some calculate the most abun-

Table 4.6. Quantification of Rodent Remains Divided by Unit from a Rock-shelter in Big Bend National Park, Texas

Rodentia Taxa	Unit S200W201		Unit S194W193	
	NISP	MNI	NISP	MNI
Cynomys (prairie dog)			1	1
Erethizon (porcupine)	2	1	2	1
Geomys (pocket gopher)			2	1
Neotoma (packrat)	13	8	29	9
Ondatra (muskrat)	1	1		
Peromyscus (mouse)	1	1		
Reithrodontomys (harvest mouse)	1	1	1	1
Sciurus (squirrel)	1	1	5	1
Sigmodon (cotton rat)	1	1	5	3
Spermophilus (ground squirrel)			2	1
Total	20	14	47	18

dant element, whereas others distinguish left from right elements or try to match different elements according to size, age, and sex of the animals.

CONCLUSION

There is a wide variety of ways to analyze archaeobiological remains. One way is not right and one way is not wrong; there are only better ways depending on particular circumstances, such as project goals, amount of and preservation of material recovered, potential research ideas and questions, and time and money constraints. It is up to the archaeologist in conjunction with the archaeobiologist to determine which methods will be most useful on the material at hand. It is essential to be flexible during analysis, because new techniques or ideas may be developed, or alternative avenues of study may be revealed. It is important to obtain the most information from the database, and ways in which to do that may not be understood at the beginning of a project but may be developed and changed as the archaeobiological material is recovered and analyzed.

5

INTEGRATION

No matter what types of archaeobiological analyses are conducted in a project, the most important aspect of interpretation is integration of the various assemblages and analyses (Sobolik 1994; Reitz et al. 1996). Integration of archaeobiological analyses is the means to obtaining the most complete picture of lifeways and paleoenvironments. Integrating studies of different assemblages can be difficult, given the variety of remains and the diverse ways archaeobiologists identify, analyze, and interpret them. In many of the case studies presented in this volume, one researcher or a number of researchers have attempted to integrate diverse archaeobiological analyses even recognizing that basic techniques, such as quantification, were not uniform across disciplines and/or even between analysts. Archaeobiological reconstructions, however, can rarely be effective, encompassing, and broad-based without integration.

For example, Crane and Carr (1994) analyzed faunal and botanical subsistence data from Cerros, a Late Preclassic Mayan community in northern Belize. They focused on data dating to before, during, and after a change at the site from an egalitarian village to a stratified society with an elite class. The authors tried to analyze and interpret their subsistence data (faunal remains by Carr, botanical remains by Crane) in the same fashion so changes in subsistence practices through time could be addressed in a comparable manner. They discussed the various quantification and comparative methods used by researchers analyzing botanical and bone remains and realized that there were few comparable methods. They had to use the ubiquity method, the only quantification method that would be comparable for both data sets. Their analyses indicated that during all stages analyzed at Cerros, people ate a large quantity and variety of marine resources as well as corn. Through time the consumption of deer, peccary, dog, turtles, and tree fruits increased, possibly due to the increase in consumption of rarer dietary items obtained for and by the elite. The elite also had a diet high in marine resources and corn, but they preferred rarer terrestrial and freshwater resources, thus accounting for the increase of those remains in site deposits through time. The authors conclude that "an integrated approach to the study of paleonutrition is essential because humans are broad-spectrum omnivores whose basic nutritional requirements

5.1. CASE STUDY: INTEGRATION OF ARCHAEOBIOLOGICAL ANALYSES TO ASSESS DIET, HEALTH, AND MIGRATORY PATTERNS IN THE LOWER PECOS REGION OF SOUTHWESTERN TEXAS

I conducted archaeobiological analyses of botanical, faunal, and paleofecal material recovered from two rock-shelters in the Lower Pecos Region of southwestern Texas. In addition, I synthesized results from previously analyzed botanical, faunal, paleofecal, and human skeletal analyses to present an integrative understanding of the paleonutrition of the prehistoric peoples of this region (Sobolik 1994). The archaeobiological remains (plants, bones, paleofeces) indicated that the population had access to a wide variety of dietary items that provided the necessary nutrients for good nutrition. Dietary staples, foods that were eaten almost on a daily basis and provided year-round nutrition, included agave, yucca, and sotol bulbs, prickly pear pads, a wide variety of rodents, and artiodactyls (deer and antelope). The nutritional content of these dietary staples provided essential energy (kcal) through fat, carbohydrates, and protein, as well as important fiber, vitamins, and minerals. The dietary array also included a wide variety of seeds and nuts from prickly pear, mesquite, yucca, acacia, hackberry, walnuts, acorns, and Mexican buckeye; evidence of pollen ingestion of grass, hackberry, cactus, and mustard by eating flowers or drinking teas (table 5.1); and ingestion of meat from fish, turtles, snakes, birds, rabbits, and rodents. Important vitamins, minerals, and trace elements were provided by eating these diverse dietary items. However, the human skeletal remains from this region indicate that although the people experienced few long-term pathological diseases, the population exhibited a high frequency of *enamel hypoplasias*. Enamel hypoplasias are growth arrest lines that appear as indentations in tooth enamel when the enamel stops growing, usually due to nutritional stress, but cessation of growth is short-lived and enamel is formed again. Therefore, the human skeletal remains indicate that some people in the population experienced short-term, nutritionally related stress. Since the stress probably wasn't caused by the quality of the diet, which had the necessary nutrients, it was probably caused by quantity of the diet or lack can be met by many combinations of a variety of plant and animal foods" (Crane and Carr 1994:77).

In another integrative analysis, I analyzed botanical, faunal, and paleofecal remains from the Lower Pecos Region of southwestern Texas, comparing that archaeobiological information to human skeletal analyses (see the case study in sidebar 5.1). The botanical, faunal, and paleofecal remains indicated that the diet of Lower Pecos populations was relatively stable and provided all the nutrients necessary for a healthy existence (Sobolik 1994). The human skeletal remains, however, indicated that there was dietary stress in the

of dietary items, possibly during the lean winter months when a wide variety of food sources were unavailable, or during weaning, a nutritionally and emotionally stressful time in a child's life.

The paleonutrition information from the Lower Pecos Region was then used to test a hypothesized seasonal round cycle for the prehistoric population (Sobolik 1996). The people occupied a territorial range, which they marked with rock art, centered on the Pecos and Devils Rivers and the Rio Grande (figure 5.1). Shafer (1986) postulated that the people followed a cyclical seasonal round based on the availability and quantity of food resources. This seasonal pattern included the desert areas and lower canyons (e.g. Frightful Cave and Conejo Shelter) during the spring to early summer, when foods such as flowers, bulbs, fruits, and plums were available. In the late summer, people moved to deeper canyon regions along the Pecos and Rio Grande (e.g., Hinds Cave) to take advantage of aquatic and upland resources. During the fall, they moved to the Devils River drainage (e.g., Baker Cave) to use acorns, walnuts, and pecans abundant in that area at that time. In the winter, when food supplies were low and available plant foods were primarily restricted to desert succulents such as yucca, agave, sotol, and prickly pear, people moved to the northern fringes (Edwards Plateau) and focused their time and energy on the acquisition of upland game.

The validity and feasibility of such a seasonal round were tested using one of the most important variables in mobility: the acquisition of resources, particularly resources related to diet. Therefore, the paleonutritional analysis of the population was essential to test the hypothesized seasonal round. The seasonal round allowed the population to maximize the food resources available in the region by following the seasonal availability and growth cycles of resources and corresponds to the nutritional information. Only through integrating many types of archaeological evidence could it be observed that populations in the Lower Pecos Region followed a seasonal mobility pattern that allowed them to maximize the dietary resources available in the environment. By maximizing seasonal resources, the population maintained a nutritionally sound and stable diet year-round.

Table 5.1.   Most Frequent Dietary Botanical Items from the Lower Pecos Region

Botanical Item	Scientific Name	Portion Used
	Fiber	
Agave	Agave lechuguilla	Leaf bases
Prickly pear	Opuntia sp.	Pads
Yucca	Yucca sp.	Leaf bases
Sotol	Dasylirion sp.	Leaf bases
Mesquite	Prosopis sp.	Pods
Onion	Allium drummondii	Bulbs
	Seeds and Nuts	
Prickly pear	Opuntia sp.	Fruit
Mesquite	Prosopis sp.	Beans
Walnut	Juglans sp.	Nuts
Acorn	Quercus sp.	Nuts
Yucca	Yucca sp.	Fruit
Acacia	Acacia sp.	Pods and beans
Hackberry	Celtis sp.	Berries
Mexican buckeye	Ugnadia speciosa	Nuts
	Pollen	
Grass	Poaceae	Achenes
Hackberry	Celtis sp.	Flowers
Sotol	Dasylirion sp.	Flowers
Agave	Agave sp.	Flowers
Yucca	Yucca sp.	Flowers
Cactus	Cactaceae	Flowers
Mustard	Brassicaceae	Flowers/achenes

From Sobolik (1994).
Note: Botanical items are presented in order from most to least frequent.

population as evidenced through enamel hypoplasias, a short-term nutritional defect. The integration of these archaeobiological data sets revealed that the population had access to necessary nutrients, but that short-term stress, most likely during weaning or during the winter months when less food was available, was a problem for this population. The paleonutritional assessment of this population was also analyzed in conjunction with a patterned seasonal round that had previously been hypothesized for them (Sobolik 1996). Therefore, in this case, the integration of archaeobiological analyses was not only conducive to nutritional and health assessments but was also used for analyses of population seasonal movement patterns. (See figure 5.3.)

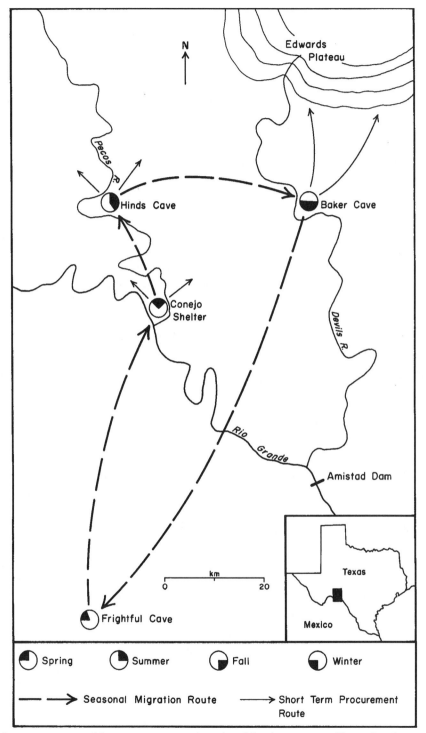

Figure 5.1. Map of the Lower Pecos Region, the Chihuahuan Desert, illustrating the seasonal round hypothesized by Shafer in 1986. (From Sobolik 1996.)

Integrative archaeobiological research does not always mean that a number of separate analyses should be combined into a comprehensive conclusion. Integration can and should involve research on other important components, such as modern species ecology, ethnography, and ethnohistory. For example, Warren (1991) demonstrates the importance of using freshwater mussels as paleoenvironmental indicators. He reviews mussel ecology and previous environmental work focusing on freshwater mussels, including habitat preferences of 133 different freshwater mussel species of the Mississippi River Basin, referencing water-body type, water depth, current velocity, and substrate composition. These preferences are given a quantitative measure that can be used in conjunction with ubiquity, MNI, and NISP and allow for statistical comparison. Warren discusses the strengths and limitations of this method and provides a number of archaeological case studies from the southeastern United States and Great Plains applying these computations and environmental interpretations.

Integrative archaeobiological work also can involve ethnographic or ethnohistoric information where feasible. For example, Snyder (1991) provides ethnohistoric, ethnographic, and archaeological evidence for the use of dogs as a dietary food item by Native Americans. Archaeological evidence of such practices was provided by faunal remains from two sites on the Great Plains. Evidence included a high frequency of cut marks on dog bones, particularly as compared to other large animals (bison and deer) that were processed at the sites. The dietary importance of dogs as food resources was assessed through a nutritional analysis of dogs as compared to other Native American foods.

In another case study involving the integration of historical information and archaeobiological analyses, Crader (1990) analyzed faunal remains found in two slave dwellings, Building "o" and the storehouse, at Monticello, Thomas Jefferson's Virginia plantation, to compare dietary practices of slaves within Monticello as compared to slaves from other habitations. Particular emphasis was placed on body part/element analysis and butchery patterns. A variety of species were identified from the assemblages, ranging from domesticated species such as pig, cattle, and sheep to wild animals such as deer, duck, and turtle. Body part representation data indicate that Building "o" dwellers obtained high-quality cuts of meat of domesticated animals (upper front limbs, upper and lower back limbs, lower back), representing a much higher frequency for quality meat than has been observed in any other analysis of slave diet (figures 5.2 and 5.3). Diet from the storehouse, however, revealed lower-quality meat part representation, what would be expected of slave quarters during this time period. The author concludes that these faunal distribution patterns may be attributed to the presence of a hierarchical system in which some slaves had higher status than others. Other hypotheses indicate that high-quality bone remains may be the result of secondary by-product reuse of leftover meat from the mansion, and that bone deposits in Building "o" merely reflect plantation garbage in general, rather than slave diet in particular.

Another important way in which to integrate archaeobiological analyses is to place them into an anthropological context. Many times archaeologists become caught up in identifying, counting, and measuring artifacts or biological remains. Their focus is on the material remains of past cultures and not on the cultures themselves.

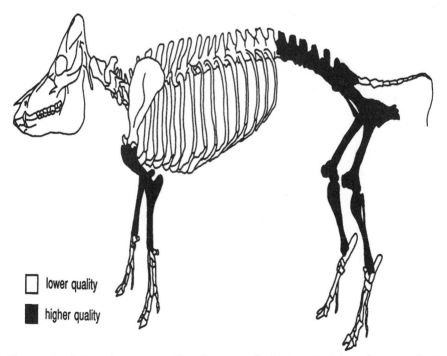

☐ lower quality

■ higher quality

Figure 5.2.   Schematic representation of meat quality in the colonial South, summarized from historical documents and archaeological information. (From Crader 1990.)

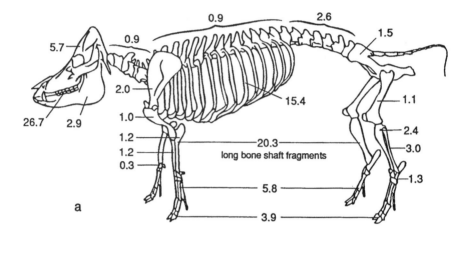

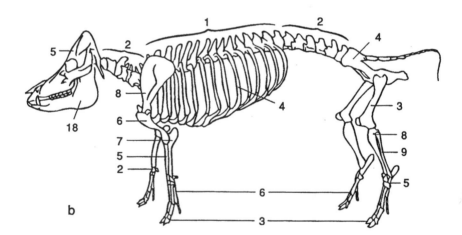

Figure 5.3. Body-part representation for pig specimens in the Building "o" assemblage: (a) number of identified specimens (N) for various body parts expressed as a percentage of the total; (b) minimum number of individuals (MNI) represented by various body parts. (From Crader 1990.)

Archaeology is a subdiscipline of anthropology, and it is important that archaeologically derived information be placed within the broader framework of cultural analysis and anthropological determination. The archaeologist and the archaeobiologist must look up from their analyses and view the bigger picture.

Kamp and Whittaker (1999) provide an excellent case study of archaeological excavation and analysis of a small Sinagua village site in Arizona (see the case study in sidebar 5.2). The authors analyze occupation, architecture, building periods, and taphonomy of the site; the types and formation of lithics, ceramics, and other artifacts; distribution and implications of botanical and faunal remains; and demography and pathology of burials. Instead of writing a typical site report detailing their analyses, identifications, and conclusions, Kamp and Whittaker provide an anthropological analysis of the people that occupied Lizard Man Village about nine hundred years ago. They compare results from their analyses to other Sinagua research, looking at stress and food shortage, subsistence strategies, social networks, and hierarchical models. Ultimately, they summarized the people's social adaptations to a marginal environment and concluded with the following:

> There were bad times when the crops failed, a child died, or your teeth hurt. There were good times: marriages, successful ceremonies, rain on the corn, a full harvest, a joke played on a friend. Your life, the lives of your ancestors, and the lives you expected for your children were no better or worse than most folk you knew, and similar to the lives of thousands in worlds beyond your ken. (Kamp and Whittaker 1999:195)

It is important to remember that people, like you and me, helped form the archaeological record, and it is their life that we are trying to reconstruct.

The focus of most CRM projects is on assessing the importance of a site before it is potentially destroyed. If a site is determined to be culturally significant, then there are two subsequent pathways. One is that the project potentially impacting the cultural site will either stop or move elsewhere. Another is that the cultural site will be excavated to retrieve the culturally significant material before it is impacted. Because the main goals of CRM projects are to assess, remove, and excavate culturally significant sites, CRM projects potentially excavate and analyze some of the most important archaeological information available today. Most archaeological excavations conducted in the United States today are CRM based. Therefore, it is

5.2. CASE STUDY: SURVIVING ADVERSITY:
THE SINAGUA OF LIZARD MAN VILLAGE

The Sinagua are a prehistoric group of southwestern people who lived in present-day central Arizona at the cultural confluence of the Anasazi, Hohokam, and Mogollon. The earliest occupation of Sinagua in the area occurs at about A.D. 600 and continues until after about A.D. 1400 during which period they seem to amalgamate with modern Hopi populations. Sunset Crater volcano erupted in the area in A.D. 1064, producing a very marginal environment for human habitation. Kamp and Whittaker (1999) analyze the adaptation of the Sinagua after the eruption of Sunset Crater in this marginal environment through the analysis of Lizard Man Village, a small Puebloan and pithouse village occupied from A.D. 1065 to 1250.

Archaeobiological analyses included plant and animal remains from the site in conjunction with analysis of human burials. Plant remains, analyzed by Andrea Hunter (Hunter et al. 1999), indicate that corn, three types of beans, and squash were the dietary staples of the population, although other plant items were also important for the diet and for other cultural needs, such as medicines, dyes, and clothing (table 5.2). Faunal remains indicated that hunting supplemented farming and gathering with rabbits and artiodactyls as the primary food animals (table 5.3). The human skeletal remains from fifteen burials indicate that the population experienced nutritional stress and poor health by the presence of high childhood mortality (47 percent); the lowest mean stature of southwestern populations (Sobolik 2002); and the presence of porotic hyperostosis, cribra orbitalia, and enamel hypoplasias.

Kamp and Whittaker conclude that the Sinagua of Lizard Man Village were a small, self-sufficient village using a mixed subsistence strategy of farming corn, beans, and squash (and possibly bug seed, amaranth, and cotton) and gathering and hunting a wider variety of resources. "Food shortages were not uncommon and existence was often precarious" (1999:185), making the existence of social networks very important for survival in this marginal environment. Social organization focused on egalitarianism and a peaceful existence while spreading populations across the landscape to maximize the use and distribution of natural resources. The Sinagua lived in relatively dispersed, nonhierarchical societies in which central authority was not present.

Of particular importance in this project is the attention Kamp and Whittaker pay to placing archaeological information within an anthropological, cultural context. They discuss what the lives of the Sinagua people would have been like, how mothers and fathers would have felt at the death of their children, and the forms of interaction that were necessary to survive in such a marginal and potentially harsh environment. They also speculate about the social interactions between people on the household, local village, and regional levels. Integration of all data sets, including archaeobiological, have allowed the researchers to frame their archaeological work within the broader anthropological context.

important that archaeologists and archaeobiologists who have access to and analyze the most numerous and culturally significant archaeological material conduct such analyses in the most integrative and synthetic way possible. It is important to go beyond the most basic goals of CRM projects and to develop integrative ways to interpret and analyze prehistoric lifeways and paleoenvironments. Only through such integration will archaeologists and archaeobiologists be able to understand the broader picture of prehistoric cultures and their implications to modern people.

Fishel (1999) provides an example of integrative analysis and interpretation of remains recovered from an archaeological site that was being severely impacted, in this case by the flooding of the Little Sioux River, in addition to being periodically pothunted by local collectors and school classes (see the case study in sidebar 5.3). Fishel is the editor of this volume in which a number of archaeologists provide a synthetic report of material remains from the large site. Most important, the information gained through excavation and analysis of the site provided extensive discussion and synthesis of prehistoric and potentially historic Oneota people. This case study provides a prime example of how data obtained from impacted archaeological sites can be integrated into comprehensive analyses to provide synthetic statements and discourse on the people at an anthropological level.

Placing archaeobiological information within an anthropological context is essential to understanding the past. Integration of archaeobiological and other archaeological remains is integral to such an assessment. Ultimately, integration has to take place at the site or regional level as the archaeobiologist works with the archaeologist (they may be the same person) to synthesize all of the information obtained from site excavation into a cohesive final statement.

I hope that after reading this book you are more familiar with the techniques used in archaeobiology and have a clearer understanding of their importance in the reconstruction of lifeways and paleoenvironments. The goals of this book were to provide you with basic procedures needed to recover and identify archaeobiological remains and to help you understand the importance of taphonomy for analysis and interpretation. In addition, I hope the case studies help you understand the importance of integrative analyses in archaeobiological research.

Table 5.2. Economic Categories of Seeds from Flotation at Lizard Man Village

Category	Pithouse Occupation				Pueblo Rooms			
	Rank	Number	Identifiable Seeds%[a]	Presence %	Rank	Number	Identifiable Seeds%	Presence %
Corn (Zea mays)	2	459	(56.7)	86	1	160	35.3	100
Starchy-seeded, possibly cultivated (Corispermum hyssopifolium, Amaranthus spp., Chenopodium spp.[b])	1	3958	86.2	100	3	56	12.4	100
Starchy-seeded noncultivated (Portulaca oleracea, Atriplex canescens, Sporobolus cryptandrus)	3	115	(14.2)	86	4	52	11.5	100
Oily-seeded noncultivated (Helianthus cf. petiolaris)	4	19	(2.3)	43	7	6	1.3	40
Domesticated, dietary/technological (Phaseolus vulgari, P. acutifolius, Gossypium hirsutum)	6	7	(1.1)	57	2	144	31.8	20
Fruits and berries (Juniperus monosperma, J. osteoperma, Coryphanta vivopara. Opuntia polycantha, Solanum spp., Arctostaphylos pringlei, Sambucus spp.)	7	6	(.7)	14	8	5	1.1	40

Continued

Table 5.2.   (continued)

Category	Pithouse Occupation				Pueblo Rooms			
	Rank	Number	Identifiable Seeds%[a]	Presence %	Rank	Number	Identifiable Seeds%	Presence %
Medicinal/dyes, low dietary (*Gutierrezia sarothrae, Polygonum amphibium, Atriplex canescens, Lepidium* spp., *Artemisia* spp.)	5	15	(1.9)	57	6	10	2.2	40
Grass and weed seeds (*Poaceae, Polygonum* spp. *Hordeum jubatum, Echinochloa crusgalli, Polygonum ramosissim*)	9	5	(.6)	29	5	18	4	80
Other	7	6	(.7)	43[c]	9	2	.4	20
Total		4590	(100)	7[c]		453	100	5[c]

Source: From Kamp and Whittaker (1999).
[a]Percentages in parentheses calculated without *Corispermum hyssopifolium.*
[b]Includes Chenopodiaceae endosperm.
[c]Total number of discrete proveniences.

Table 5.3. Major Faunal Groups by Room at Lizard Man Village

Taxa	Lagomorphs				Artiodactyls				Large Rodents[a]				Small Rodents[b]			
	NISP	MNI	%NISP	%MNI	NISP	MNI	%NISP	%MNI	NISP	MNI	%NISP	%MNI	NISP	MNI	%NISP	%MNI
Room 3	416	25	71	39	12	4	2	6	70	10	12	15	51	15	9	23
Room 4	185	10	62	26	44	5	15	13	24	4	8	10	35	10	12	26
Room 6	195	19	66	42	13	3	4	7	53	9	18	20	21	8	7	18
Room 7	82	12	47	32	5	3	3	8	50	7	29	19	25	9	14	24
Room 8	63	9	57	35	5	3	3	12	29	5	26	19	8	5	7	19
Room 11	85	8	50	24	4	2	2	6	32	4	19	12	43	14	25	42
Room 15	328	18	75	35	18	3	4	6	57	10	13	20	20	8	5	16
Room 16	194	19	80	49	6	3	3	8	22	4	9	10	14	7	6	18
Room 17	91	6	76	35	0	0	0	0	8	2	7	12	9	6	8	35
Room 19	157	13	65	37	36	3	15	9	25	7	10	20	12	8	5	23
"Room" 24	225	13	70	28	17	4	5	9	19	5	6	11	45	16	14	35
Total	2021	152	67	35	160	33	5	8	389	67	13	16	283	108	9	25

Continued

Table 5.3. *(continued)*

Taxa	Carnivores[c]				Birds				Reptiles		Room Totals	
	NISP	MNI	% NISP	% MNI	NISP	MNI	% NISP	% MNI	NISP	MNI	NISP	MNI
Room 3	4	1	1	2	37	8	6	12	0	0	590	65
Room 4	2	1	1	3	8	8	3	21	2	1	300	39
Room 6	8	2	3	4	4	4	1	9	0	0	294	45
Room 7	2	1	1	3	10	5	6	14	0	0	174	37
Room 8	1	1	1	4	4	2	4	8	1	1	111	26
Room 11	1	1	1	3	5	4	3	12	0	0	170	33
Room 15	5	3	1	6	10	8	2	16	1	1	439	51
Room 16	0	0	0	0	6	5	3	13	1	1	243	39
Room 17	8	1	7	6	4	2	3	12	0	0	120	17
Room 19	6	2	3	6	4	2	2	6	0	0	240	35
"Room" 24	5	2	2	4	7	5	2	11	3	1	321	46
Total	42	15	1	4	99	53	3	12	8	5	3002	433

Source: From Kamp and Whittaker (1999).
[a]Large rodents include prairie dogs, rock squirrels, and other squirrel-size specimens.
[b]Small rodents includes all other rodents.
[c]Carnivores include canids and mustelids.

5.3. CASE STUDY: BISON HUNTERS OF THE WESTERN PRAIRIES: ARCHAEOLOGICAL INVESTIGATIONS AT A LARGE ONEOTA SITE IN WESTERN IOWA

Excavations at the Dixon Site (13WD8) in Woodbury County, Iowa, were conducted because the large Oneota site had been extensively eroded due to expansion of the Little Sioux River during the 1993 flood, and cultural materials, including human bones, were eroding from the riverbank. In addition, the site had been and continues to be pothunted by local collectors as well as a local high school teacher who took classes out to the site to dig. Results from excavations and analysis of the cultural remains were synthesized and edited by Fishel (1999), although a number of archaeologists helped with material analysis and interpretation.

The research design included excavation and analysis of fifty pit features that were exposed and eroding from the western cutbank wall. Four research goals and questions were addressed through analysis and interpretation of the remains (table 5.4). Archaeobiological remains were useful for helping to answer all of these questions but were particularly useful to answer questions 1, 2, and 4.

The archaeobiological remains included large quantities of botanical remains from flotation, as well as faunal remains and human skeletal material. The Dausman Model A Flote-Tech Method was used for floating 972 liters of sediment from thirty-six features. The flotation samples varied from 1 liter to 53.5 liters depending on how large the features were. Botanical remains were separated into size grades of greater and lesser than two millimeters. Table 5.5 summarizes the results from flotation, including the density, weights, and counts of botanical items. Maize constituted a large part of the diet, and squash rinds were frequently observed, although other cultivated crops are not found, such as the eastern cultigens of maygrass, knotweed, and large chenopods. Wood, mainly oak and elm, was chosen from the bottomlands and along stream terraces.

Faunal remains included 3,368 bones, of which 13.7 percent were identifiable due to fragmentation of the assemblage because of breakage and processing by the Oneota (table 5.6). Bone from bison was the highest percentage of identifiable bone. A minimum of thirteen individuals were excavated revealing a preponderance of males taken as food items and for other uses, such as scapula hoes, bone tools, and furs. Deer, elk, canids, beaver, and pocket gopher were also frequently used by prehistoric Oneota. Of interest are 639 freshwater mussel shell fragments representing nine different species. Freshwater mussel was used as a food resource, for tempering ceramics, and for tool manufacture.

In response to the original four research design questions (table 5.4), some have been partially answered, and others remain to be answered. Archaeobiological remains helped realize the large extent of the Dixon site and that there may exist upwards of 9,800 additional pit features, many with human remains. Botanical processing areas were identifiable around feature 16, and more probably exist. Analysis of flotation remains indicated that Oneota diet focused on maize with little exploitation of native cultigens or wild plant foods. Hunting of bison probably took place in long-distance communal forays, indicating that bison were not frequent in the Dixon site area in the past. The Oneota of the region are better understood and it is realized that Oneota postdate Mill Creek groups in the region and that the Dixon site Oneota are one of the earliest manifestations of Oneota in the area, although it is not understood where the Oneota peoples came from. Integrative archaeological, archaeobiological, and historical analysis work together to answer research design questions.

Table 5.4.   Research Design Goals and Questions for Excavation and Analysis of the Dixon Site, Western Iowa

Research Design Goals and Questions	Steps Undertaken and Archaeological Material Used
1. How did Oneota peoples first come to occupy northwest Iowa, and what were the initial settlements like?	Analysis of features and associated material remains
2. What was the relationship, if any, between Oneota groups and other prehistoric peoples who inhabited northwest Iowa?	Collect numerous radiocarbon samples from various contexts
3. What kinds of cultural remains exist at the Dixon site, and how are those remains spatially distributed? What measures can be taken to help safeguard the site against continued damage from erosion?	Magnetometer and soil resistivity surveys, topographic maps, and geomorphological assessments
4. What do the cultural remains at the Dixon site tell us about the lifeways of Oneota groups living in northwest Iowa?	Flotation samples from features, macrofaunal remains, lithics, ceramics, and human remains

Source: From Fishel (1999).

Table 5.5. Archaeobotanical Indices from Feature Flotation Samples at the Dixon Site, Northwest Iowa

Taxon	Common Name	MNI	Count	Percentage	Weight (g)	Percentage
Bison bison	bison	13	179	5.32	7,333.50	63.70
Odocoileus virginianus	white-tailed deer	1	1	.03	491.50	4.27
Odocoileus sp.	deer	6	96	2.82	802.30	6.97
Canis sp.	dog or wolf	3	64	1.90	259.70	2.26
Cervus canadensis	elk	2	10	.30	61.70	.54
Castor canadensis	beaver	1	5	.15	15.70	.14
Scalopus aquaticus	Eastern mole	1	3	.09	.90	.00
Rattus sp.	rat	1	2	.06	.70	.00
Procyon lotor	raccoon	1	2	.06	11.50	.10
Sus scrofa	pig	1	8	.24	13.10	.11
Geomys bursarius		1	2	.06	1.10	.00
Sylvilagus sp.	rabbit	1	1	.39	.90	.01
Sciurid		1	13	.39	1.50	.01
Turtle		1	3	.09	8.60	.07
Frog		1	3	.09	.20	.00
Fish (*Ictalurus* sp.)		6	64	1.90	16.00	.14
Tetraoninae sp.		1	2	.06	.90	.00
Anseriformes sp.		1	1	.03	.50	.00
Cf. *Icterinae* sp.		1	1	.03	.50	.00
Cf. *Anatidae* sp.		1	1	.03	.40	.00
Unidentified Aves		1	1	.03	.50	.00
Burned unidentifiable			342	1.15	192.70	1.67
Unidentifiable			2,564	76.13	2,297.20	19.95
Total			3,368		11,511.60	

Source: From Fishel (1999).

Table 5.6. Excavated Botanical Remains from the Dixon Site, Northwest Iowa

Sum raw charcoal wt (g), nonstandardized samples	125.405
Sum charcoal wt (g), standardized 10-liter samples	120.791
Average wt (g) charcoal/sample in standardized samples	1.267
Average wt (g) charcoal/feature in standardized samples	3.485
Sum flotation volumes (liters)	972.000
Average flotation volume (liters) per sample	9.818
Average flotation volume (liters) per feature	27.000
Sum raw counts, carbonized seeds, all sizes	3,027.000
Estimated sum seeds, all sizes, in standardized 10-liter samples	3,194.011
Seeds/g of charcoal in standardized samples	49.649
Seeds/g of charcoal by raw count	24.138
Sum raw counts, charcoal >2 mm	7,165.000
Sum standardized sample counts, charcoal <2 mm	6,685.088

	Count/10liters Charcoal >2mm	Frequency (%)	Feature Ubiquity	Sample Ubiquity
Nutshell	(59.967)	(0.9)	97.2	78.8
Hazelnut	14.755	0.22	11.1	19.4
Juglandaceae	4.437	0.07	75.0	57.6
Black walnut	40.775	0.61	80.6	56.6
Wood	5,202.403	77.82	100.0	100.0
Bark	69.762	1.04	58.3	47.5
Grass stems	57.887	0.87	58.3	33.3
Tubers/rhizomes	6.47	0.10	25.0	10.1
Seeds: Freq = >2mm; Ubiqu = >0.5	20.843	0.31	97.2	97.0
Squash (cucurbit) rind	10.454	0.16	83.3	76.8
Maize kernels	867.098	12.97	83.3	69.7
Maize cupules	307.943	4.61	83.3	66.7
Maize embryos	6.697	0.10	36.1	16.2
Unknowns	45.583	0.68	66.7	49.5

Source: From Fishel (1999).
[a]Frequency: % charcoal >2mm, standardized per sample to 10-liter volumes before computation. Sample ubiquity is a tabulation of the presence of a charcoal type in any sample, any fraction size, expressed as a percentage of total samples (*n*=99). Feature ubiquity is a tabulation of the presence of a charcoal type in a feature, any sample, any fraction, expressed as a percentage of total number of features (*n*=36).

 **REFERENCES**

Appleyard, H. M., and A. B. Wildman
   1970   Fibres of Archaeological Interest: Their Examination and Identification. In *Science in Archaeology: A Survey of Progress and Research*, D. Brothwell and E. Higgs (eds.), pp. 624–33. Praeger, New York.

Asch Sidell, Nancy
   1999   Prehistoric Plant Use in Maine: Paleoindian to Contact Period. In *Current Northeast Paleoethnobotany*, John Hart (ed). New York State Museum, Albany.

Barrows, David
   1900   *The Ethno-botany of the Coahuilla Indians of Southern California*. Ph.D. dissertation, University of Chicago, Chicago.

Birks, H. J. B.
   1981   Late Wisconsin Vegetational and Climatic History at Kylen Lake, Northeastern Minnesota. *Quaternary Research* 16:322–55.

Bocek, Barbara
   1986   Rodent Ecology and Burrowing Behavior: Predicted Effects on Archaeological Site Formation. *American Antiquity* 51(3):589–603.

Bokonyi, S.
   1970   A New Method for the Determination of the Minimum Number of Individuals in Animal Bone Material. *American Journal of Archaeology* 74:291–92.

Bozell, John R.
   1991   Fauna from the Hulme Site and Comments on Central Plains Tradition Subsistence Variability. *Plains Anthropologist* 36(136):229–53.

119

Braidwood, L. S., and R. J. Braidwood (eds.)
   1982   *Prehistoric Village Archaeology in South-Eastern Turkey: The Eight Millennium B.C. Site at Cayönü: Its Chipped and Ground Stone Industries and Faunal Remains.* British Archaeological Reports International Series 138. Oxford University Press, Oxford.
Bryant, Vaughn M. Jr.
   1966   Pollen Analysis of the Devil's Mouth Site. In *A Preliminary Study of the Paleoecology of the AMISTAD Reservoir.* A Report of Research under the Auspices of the National Science Foundation (GS-667). National Science Foundation, Washington, D.C.
Bryant, Vaughn M. Jr., Richard G. Holloway, John G. Jones, and David L. Carlson
   1994   Pollen Preservation in Alkaline Soils of the American Southwest. In *Sedimentation of Organic Particles,* Alfred Traverse (ed.), pp. 47–58. Cambridge University Press, Cambridge.
Bryant, Vaughn M. Jr., and Don P. Morris
   1986   Uses of Ceramic Vessels and Grinding Implements: The Pollen Evidence. In *Archeological Investigations at Antelope House,* D. P. Morris (ed.), pp. 489–500. National Park Service, Washington, D.C.
Bryant, Vaughn M. Jr. and Glendon H. Weir
   1986   Pollen Analysis of Floor Sediment Samples: A Guide to Room Use. In *Archeological Investigations at Antelope House,* D. P. Morris (ed.), pp. 58–71. National Park Service, Washington, D.C.
Bunn, H. T.
   1991   A Taphonomic Perspective on the Archaeology of Human Origins. *Annual Review of Anthropology* 20:433–67.
Cannon, Debbi Yee
   1987   *Marine Fish Osteology: A Manual for Archaeologists.* Archaeology Press, Simon Fraser University, Burnaby, Canada.
Carbone, Victor A., and Bennie C. Keel
   1985   Preservation of Plant and Animal Remains. In *The Analysis of Prehistoric Diets,* Robert I. Gilbert Jr. and James H. Mielke (eds.), pp. 1–20. Academic Press, Orlando, Florida.
Casteel, Richard W.
   1976   *Fish Remains in Archaeology and Paleoenvironmental Studies.* Academic Press, New York.
   1977   Characterization of Faunal Assemblages and the Minimum Number of Individuals Determined from Paired Elements: Continuing Problems in Archaeology. *Journal of Archaeological Science* 4:125–34.
   1978   Faunal Assemblages and the "Weigemethode" or Weight Method. *Journal of Field Archaeology* 5:71–77.

Catling, Dorothy, and John Grayson
  1982 *Identification of Vegetable Fibres.* Chapman and Hall, London.
Claasen, Cheryl
  1998 *Shells.* Cambridge University Press, Cambridge.
Cooke, Richard G., Lynette Norr, and Dolores R. Piperno
  1996 Native Americans and the Panamanian Landscape. In *Case Studies in Environmental Archaeology*, E. J. Reitz, L. A. Newsom, and S. J. Scudder (eds.), pp. 103–26. Plenum Press, New York.
Core, H. A., W. A. Cote, and A. C. Day
  1979 *Wood Structure and Identification.* Syracuse University Press, Syracuse, New York.
Corner, Edred J. H.
  1976 *The Seeds of Dicotyledons.* Cambridge University Press, Cambridge.
Crader, Diana C.
  1990 Slave Diet at Monticello. *American Antiquity* 55(4):690–717.
Crane, Cathy J., and H. Sorayya Carr
  1994 The Integration and Quantification of Economic Data from a Late Preclassic Maya Community in Belize. In *Paleonutrition: Diet, Health, and Nutrition in Prehistory*, K. D. Sobolik (ed.), pp. 66–79, Center for Archaeological Investigations Occasional Paper No. 22. Southern Illinois University, Carbondale.
Dimbleby, Geoffrey W.
  1978 *Plants and Archaeology.* Humanities Press, Atlantic Highlands, New Jersey.
  1985 *The Palynology of Archaeological Sites.* Academic Press, New York.
Dobbs, Clark A., Craig Johnson, Kathryn Parker, and Terrance Martin
  1993 *20SA1034: A Late-Prehistoric Site on the Flint River in the Saginaw Valley, Michigan.* Reports of Investigations No. 229. Institute for Minnesota Archaeology, Minneapolis.
Efremov, J. A.
  1940 Taphonomy: A New Branch of Paleontology. *Pan-American Geologist* 74:81–93.
Faegri, Knut, and John Iversen
  1964 *Textbook of Pollen Analysis.* Hafner, New York.
Fewkes, J. W.
  1896 Pacific Coast Shell from Prehistoric Tusayan Pueblos. *American Anthropologist* 9(11):359–67.
Fishel, Richard L. (ed.)
  1999 *Bison Hunters of the Western Prairies: Archaeological Investigations at the Dixon Site (13WD8), Woodbury County, Iowa.* Report No. 21. Office of the State Archaeologist, University of Iowa, Iowa City.

Ford, Richard I.
 1979 Paleoethnobotany in American Archaeology. In *Advances in Archaeological Method and Theory, Vol. 2*, Michael B. Schiffer (ed.), pp. 285–336. Academic Press, New York.

Fry, Gary F.
 1985 Analysis of Fecal Material. In *The Analysis of Prehistoric Diets*, R. I. Gilbert Jr. and J. H. Mielke (eds.), pp. 127–54, Academic Press, Orlando, Florida.

Gilbert, A. S., and B. H. Singer
 1982 Reassessing Zooarchaeological Quantification. *World Archaeology* 14:21–40.

Gilbert, B. Miles
 1990 *Mammalian Osteology*. Missouri Archaeological Society, Columbia.

Gilbert, B. Miles, Larry D. Martin, and Howard G. Savage
 1985 *Avian Osteology*. Modern Printing, Laramie, Wyoming.

Gilmore, Melvin
 1919 *Uses of Plants by the Indians of the Missouri River Region*. Thirty-second Annual Report of the Bureau of American Ethnology. Smithsonian Institution Press, Washington, D.C. Reprinted 1977, University of Nebraska Press, Lincoln.
 1932 The Ethnobotanical Laboratory at the University of Michigan. *Occasional Contributions from the Museum of Anthropology of the University of Michigan*, No. 1. University of Michigan Press, Ann Arbor.

Grayson, Donald K.
 1984 *Quantitative Zooarchaeology: Topics in the Analysis of Archaeological Faunas*. Academic Press, New York.

Greig, James
 1989 *Archaeobotany*. Handbooks for Archaeologists, No. 4. European Science Foundation, Strasbourg.

Gremillion, Kristen J., and Kristin D. Sobolik
 1996 Dietary Variability among Prehistoric Forager-Farmers of Eastern North America. *Current Anthropology* 37(3):529–39.

Gunn, C. R., J. V. Dennis, and P. J. Paradine
 1976 *World Guide to Tropical Drift Seeds and Fruits*. New York Times Book Company, Quadrangle, New York.

Harshberger, J. W.
 1896 The Purpose of Ethnobotany. *American Antiquarian* 17(2):73–81.

Hastorf, C. A., and V. S. Popper (eds.)
 1988 *Current Paleoethnobotany: Analytical Methods and Cultural Interpretations of Archaeological Plant Remains*. University of Chicago Press, Chicago.

Helbaek, Hans
1969 Plant Collecting, Dry-Farming, and Irrigation Agriculture in Prehistoric Deh Luran. In *Prehistory and Human Ecology of the Deh Luran Plain*, F. Hole, K. Flannery, and J. Neely (eds.), pp. 389–426, Memoirs No. 1, Museum of Anthropology, University of Michigan. University of Michigan, Ann Arbor.

Hoffman, Rob, and Christopher Hays
1987 The Eastern Wood Rat (*Neotoma floridana*) as a Taphonomic Factor in Archaeological Sites. *Journal of Archaeological Science* 14:325–37.

Holloway, Richard G., and Vaughn M. Bryant Jr.
1985 Late-Quaternary Pollen Records and Vegetational History of the Great Lakes Region: United States and Canada. In *Pollen Records of Late-Quaternary North American Sediments*, V. M. Bryant Jr. and R. G. Holloway (eds.), pp. 207–45. American Association of Stratigraphic Palynologists, Dallas.

Hudson, Jean
1993 The Impacts of Domestic Dogs on Bone in Forager Camps. In *From Bones to Behavior: Ethnoarchaeological and Experimental Contributions to the Interpretation of Faunal Remains*, Jean Hudson (ed.), pp. 301–23, Center for Archaeological Investigations Occasional Paper No. 21. Southern Illinois University Press, Carbondale.

Hunter, Andrea, Kathryn Kamp, and John Whittaker
1999 Plant Use. In *Surviving Adversity: The Sinaqua of Lizard Man Village*, Kathryn Kamp and John Whittaker (eds.), pp. 139–51. University of Utah Anthropological Papers, No. 120. University of Utah Press, Salt Lake City.

Izumi, S., and T. Sono
1963 *Andes 2: Excavations at Kotosh, Peru, 1960*. Kadokawa, Tokyo.

Johnson, LeRoy Jr.
1964 *The Devil's Mouth Site: A Stratified Campsite at Amistad Reservoir, Val Verde County, Texas*. Archeology Series No. 6, Department of Anthropology, University of Texas. University of Texas Press, Austin.

Jones, Gretchen D., Vaughn M. Bryant Jr., Meredith Hoag Lieux, Stanley D. Jones, and Pete D. Lingren
1995 *Pollen of the Southeastern United States: with emphasis on melissopalynology and entomopalynology*. American Association of Stratigraphic Palynologists Foundation, AASP Contribution Series No. 30.

Jones, John G., and Vaughn M. Bryant Jr.
1992 Phytolith Taxonomy in Selected Species of Texas Cacti. In *Phytolith Systematics: Emerging Issues*, G. Rapp Jr. and S. C. Mulholland (eds.), pp. 215–38, Plenum Press, New York.

Jones, Kevin T.
1993   The Archaeological Structure of a Short-Term Camp. In *From Bones to Behavior: Ethnoarchaeological and Experimental Contributions to the Interpretation of Faunal Remains*, Jean Hudson (ed.), pp. 101–14, Center for Archaeological Investigations Occasional Paper No. 21. Southern Illinois University Press, Carbondale.

Jones, Volney H.
1936   Vegetable Remains of Newt Kash Hollosh Shelter. In *Rockshelters in Menefee County, Kentucky*, W. S. Webb and W. D. Funkhouser (eds.), University of Kentucky Reports in Archaeology and Anthropology 3:147–65.

Kamp, Kathryn A., and John C. Whittaker
1999   *Surviving Adversity: The Sinaqua of Lizard Man Village*. University of Utah Anthropological Papers, No. 120. University of Utah Press, Salt Lake City.

Kaplan, Lawrence, Mary B. Smith, and Lesley Sneddon
1990   The Boylston Street Fishweir: Revisited. *Economic Botany* 44(4):516–28.

Kapp, Ronald O.
1969   *How to Know Pollen and Spores*. Brown, Dubuque, Iowa.

Klein, Richard G., and Kathryn Cruz-Uribe
1984   *The Analysis of Animal Bones from Archeological Sites*. University of Chicago Press, Chicago.

Krantz, G. S.
1968   A New Method of Counting Mammal Bones. *American Journal of Archaeology* 72:286–88.

Labadie, Joe
1994   *Amistad National Recreation Area Cultural Resources Study*. National Park Service, U.S. Department of the Interior, Washington, D.C.

Lawrence, Barbara
1957   Zoology. In *The Identification of Non-artifactual Archaeological Materials*, W. W. Taylor (ed.), pp. 41–42, National Academy of Science–Natural Resource Council Publication 565. National Academy of Science and Natural Resource Council, Washington, D.C.

Leney, Lawrence, and R. W. Casteel
1975   Simplified Procedure for Examining Charcoal Specimens for Identification. *Journal of Archaeological Science* 2:153–59.

Linse, Angela R.
1992   Is Bone Safe in a Shell Midden? In *Deciphering a Shell Midden*, Julie Stein (ed.), pp. 327–46, Academic Press, New York.

Loomis, F. B., and D. B. Young
　1912　Shell Heaps of Maine. *American Journal of Science* 34(199):17–42.

Lyman, R. Lee
　1979　Available Meta from Faunal Remains: A Consideration of Techniques. *American Antiquity* 44:536–46.
　1984　Bone Density and Differential Survivorship of Fossil Classes. *Journal of Anthropological Archaeology* 3:259–99.
　1994　*Vertebrate Taphonomy.* Cambridge University Press, New York.

Mack, Richard N., and Vaughn M. Bryant Jr.
　1974　Modern Pollen Spectra from the Columbia Basin, Washington. *Northwest Science* 48(3):183–94.

Martin, A. C., and W. D. Barkley
　1961　*Seed Identification Manual.* University of California Press, Berkeley.

McInnis, Heather
　1999　*Subsistence and Maritime Adaptations at Quebrada Jaguay, Camaná, Peru: A Faunal Analysis.* M.S. thesis, Institute for Quaternary and Climate Studies, University of Maine, Orono.

Miller, Naomi F., and Tristine Smart
　1984　Intentional Burning of Dung as Fuel: A Mechanism for the Incorporation of Charred Seeds into the Archeological Record. *Journal of Ethnobiology* 4(1):15–28.

Montgomery, F. H.
　1977　*Seeds and Fruits of Plants of Eastern Canada and Northeastern United States.* University of Toronto Press, Toronto.

O'Connell, J. F., and K. Hawkes
　1988　Hadza Hunting, Butchering, and Bone Transport and Their Archaeological Implications. *Journal of Anthropological Research* 44(2):113–61.

O'Hara, Sara L., F. Alayne Street-Perrott, and Timothy P. Burt
　1993　Accelerated Soil Erosion around a Mexican Highland Lake Caused by Prehispanic Agriculture. *Nature* 362:48–51.

Olsen, Stanley J.
　1968　*Fish, Amphibian and Reptile Remains from Archaeological Sites.* Papers of the Peabody Museum of Archaeology and Ethnology, Harvard University, Volume 55, No. 2, Peabody Museum, Cambridge, Massachusetts.

Parker, Kathryn E.
　1996　Three Corn Kernels and a Hill of Beans: The Evidence for Prehistoric Horticulture in Michigan. In *Investigating the Archaeological Record of the Great Lakes State: Essays in Honor of Elizabeth Baldwin Garland,* Margetet B. Holman, Janet G.

Brashler, and Kathryn E. Parker (eds.), pp. 307-339. Western Michigan University, Kalamazoo.

Pearsall, Deborah M.
  1989   *Paleoethnobotany: A Handbook of Procedures.* Academic Press, New York.
  2000   *Paleoethnobotany: A Handbook of Procedures,* 2nd ed. Academic Press, San Diego.

Piperno, Dolores R.
  1988   *Phytolith Analysis: An Archaeological and Geological Perspective.* Academic Press, New York.

Piperno, Dolores R., and Deborah M. Pearsall (eds.)
  1993   *Current Research in Phytolith Analysis: Applications in Archaeology and Paleoecology.* MASCA Research Papers in Science and Archaeology, Vol. 10, University Museum of Archaeology and Anthropology. University of Pennsylvania, Philadelphia.

Poinar, Hendrik N., Melanie Kuch, Kristin D. Sobolik, Ian Barnes, Artur B. Stankiewicz, Tomasz Kuder, W. Geofferey Spaulding, Vaughn M. Bryant Jr., Alan Cooper, and Svante Pääbo
  2001   A Molecular Analysis of Dietary Diversity for Three Archaic Native Americans. *Proceedings of the National Academy of Sciences* 98(8):4317–22.

Rapp, George Jr., and Susan C. Mulholland (eds.)
  1992   *Phytolith Systematics: Emerging Issues.* Plenum Press, New York.

Reitz, Elizabeth J., Lee A. Newsom, and Sylvia J. Scudder (eds.)
  1996   *Case Studies in Environmental Archaeology.* Plenum Press, New York.

Reitz, Elizabeth J., and Elizabeth Wing
  1999   *Zooarchaeology.* Cambridge University Press, New York.

Robison, N.D.
  1978   Zooarchaeology: Its History and Development. *Tennessee Anthropological Association Miscellaneous Paper* 2:1–22 (Knoxville).

Sandweiss, Daniel H., Heather McInnis, Richard L. Burger, Asuncion Cano, Bernardino Ojeda, Rolando Paredes, Maria del Carmen Sandweiss, and Michael D. Glascock
  1998   Quebrada Jaguay: Early South American Maritime Adaptations. *Science* 281:1830–32.

Sanger, David
  1996   Testing the Models: Hunter-Gatherer Use of Space in the Gulf of Maine, USA. *World Archaeology* 27(3):512–26.

Shafer, Harry J., and Jim Zintgraff
  1986   *Ancient Texans: Rock Art and Lifeways along the Lower Pecos.* Texas Monthly Press, Austin.

Shaffer, Brian
1992   Quarter-inch Screening: Understanding Biases in Recovery of Vertebrate Faunal Remains. *American Antiquity* 57(1):129–36.
Shipman, Pat
1986   Scavenging or Hunting in Early Hominids: Theoretical Framework and Test. *American Anthropologist* 88:27–43.
Smith, Bruce
1975   *Middle Mississippi Exploitation of Animal Populations*. University of Michigan Museum of Anthropology, Anthropological Papers 57. University of Michigan Press, Ann Arbor.
Snyder, Lynn M.
1991   Barking Mutton: Ethnohistoric and Ethnographic, Archaeological, and Nutritional Evidence Pertaining to the Dog as a Native American Food Resource on the Plains. In *Beamers, Bobwhites, and Blue-Points: Tributes to the Career of Paul W. Parmalee*, J. R. Purdue, W. E. Klippel, and B. W. Styles (eds.), pp. 359–78, Illinois State Museum Scientific Papers, Vol. 23. Illinois State Museum, Springfield.
Sobolik, Kristin D.
1988   *The Prehistoric Diet and Subsistence of the Lower Pecos Region, as Reflected in Coprolites from Baker Cave, Val Verde County, Texas*. Studies in Archeology No. 7, Texas Archeological Research Laboratory. University of Texas Press, Austin.
1993   Direct Evidence for the Importance of Small Animals to Prehistoric Diets: A Review of Coprolite Studies. *North American Archaeologist* 14(3):227–44.
1994   In *Paleonutrition: the Diet and Health of Prehistoric Americans*. Center for Archaeological Investigations, Southern Illinois University Occasional Paper No. 22, Carbondale.
1996   Nutritional Constraints and Mobility Patterns of Hunter-Gatherers in the Northern Chihuahuan Desert. In *Case Studies in Environmental Archaeology*, Elizabeth J. Reitz, Lee A. Newsom, and Sylvia J. Scudder (eds.), pp. 195–214. Plenum Press, New York.
2002   Children's Health in the Prehistoric Southwest. In: *Children in the Prehistoric Puebloan Southwest*, Kathryn Kamp (ed.), pp. 125–151, The University of Utah Press, Salt Lake City.
Sobolik, Kristin D., Kristen J. Gremillion, Patricia L. Whitten, and Patty Jo Watson
1996   Technical Note: Sex Determination of Prehistoric Human Paleofeces. *American Journal of Physical Anthropology* 101:283–90.
Sobolik, Kristin D., and D. Gentry Steele
1996   *A Turtle Atlas to Facilitate Archaeological Identifications*. Mammoth Site of Hot Springs, South Dakota.

Sobolik, Kristin D., and Richard Will
    2000   Calcined Turtle Bones from the Little Ossipee North Site in Southwestern Maine. *Archaeology of Eastern North America* 28:15–28.
Sobolik, Kristin D., Laurie S. Zimmerman, and Brooke Manross Guilfoyl
    1997   Indoor versus Outdoor Firepit Usage: A Case Study from the Mimbres. *Kiva* 62(3):283–300.
Spiess, Arthur, Kristin D. Sobolik, Dinah Crader, Richard Will, and John Mosher
    1998   Cod, Clams, and Roast Deer: Prehistoric Dining on Indiantown. In *Indiantown Island Archaeological Project*, Deborah B. Wilson (ed.), submitted to Boothbay Region Land Trust, Inc., Boothbay Harbor, Maine.
Struever, Stuart
    1968   Flotation Techniques for the Recovery of Small-scale Archaeological Remains. *American Antiquity* 33:353–62.
Sutton, Mark Q.
    1994   Indirect Evidence in Paleonutrition Studies. In *Paleonutrition: Diet, Health, and Nutrition in Prehistory*, K. D. Sobolik (ed.), pp. 98–114, Center for Archaeological Investigations Occasional Paper No. 22. Southern Illinois University Press, Carbondale.
Sutton, Mark Q., Minnie Malik, and Andrew Ogram
    1996   Experiments on the Determination of Gender from Coprolites by DNA Analysis. *Journal of Archaeological Science* 23:263–67.
Thomas, D. H.
    1969   Great Basin Hunting Patterns: A Quantitative Method for Treating Faunal Remains. *American Antiquity* 34:392–401.
Voorhies, V. R.
    1969   Taphonomy and Population Dynamics of the Early Pliocene Vertebrate Fauna, Knox County, Nebraska. *University of Wyoming Contributions in Geology Special Papers*, 1:1–69.
Wagner, Gail E.
    1996   Feast or Famine? Seasonal Diet at a Fort Ancient Community. In *Case Studies in Environmental Archaeology*, E. J. Reitz, L. A. Newsom, and S. J. Scudder (eds.), pp. 255–72. Plenum Press, New York.
Warren, Robert E.
    1991   Freshwater Mussels as Paleoenvironmental Indicators: A Quantitative Approach to Assemblage Analysis. In *Beamers, Bobwhites, and Blue-Points: Tributes to the Career of Paul W. Parmalee*, J. R. Purdue, W. E. Klippel, and B. W. Styles (eds.), pp. 23–66, Illinois State Museum Scientific Papers, Vol. 23. Illinois State Museum, Springfield.

Watson, Patty Jo
  1976   In Pursuit of Prehistoric Subsistence: A Comparative Account of Some Contemporary Flotation Techniques. *Midcontinental Journal of Archaeology* 1(1):77–100.
Western, A. Cecilia
  1970   Wood and Charcoal in Archaeology. In *Science in Archaeology: A Survey of Progress and Research*, D. Brothwell and E. Higgs (eds.), pp. 178–87. Praeger, New York.
White, T. E.
  1953   A Method of Calculating the Dietary Percentage of Various Food Animals Utilized by Aboriginal Peoples. *American Antiquity* 18(4):396–98.
Will, Richard T., and James Clark
  1996   Stone Artifact Movement on Impoundment Shorelines: A Case Study from Maine. *American Antiquity* 61(3):499–519.
Will, Richard, James Clark, and Edward Moore
  1996   *Phase III Archaeological Data Recovery at the Little Ossipee North Site (7.7) Bonny Eagle Project (FERC #2529), Cumberland County, Maine.* Prepared for Central Maine Power Company, Augusta, Maine.
Winters, Howard D.
  1969   *The Riverton Culture: A Second Millennium Occupation in the Central Wabash Valley.* Illinois State Museum Report of Investigations No. 13. Illinois State Museum, Springfield.

INDEX

ABOUT THE AUTHOR AND SERIES EDITORS

Kristin D. Sobolik received her B.S. in biology from the University of Iowa (1986), M.A. in anthropology from Texas A&M University (1988), and Ph.D. (with a dissertation entitled "Paleonutrition of the Lower Pecos Region") from Texas A&M (1991). She held a postdoctoral fellowship position at Southern Illinois University, where she edited *Paleonutrition: The Diet and Health of Prehistoric Americans*, and then joined the faculty at the University of Maine in the Anthropology Department and the Institute for Quaternary and Climate Studies, where she is an associate professor and director of the Zooarchaeology Laboratory.

Kristin has conducted archaeological work in the desertic regions of North America focusing in the Lower Pecos and Big Bend regions of Texas as well as throughout the southwestern United States. In addition, she has conducted research in Maine and at Mammoth Cave, Kentucky. Her recent research has centered on analyzing paleonutrition and possible prehistoric sex differences using DNA and hormonal content from paleofeces. She is also interested in the zoogeography of extinct and extant animal species, children's health in the prehistoric Southwest, and evidence for prehistoric cannibalism. She lives with her husband, Scott, a writer and English instructor, and their four beautiful children ranging in age from 1 to 12.

Larry J. Zimmerman is the head of the Archaeology Department of the Minnesota Historical Society. He served as an adjunct professor of anthropology and visiting professor of American Indian and native studies at the University of Iowa from 1996 to 2002 and as chair of the American Indian and Native Studies Program from 1998 to 2001. He earned his Ph.D. in anthropology at the University of Kansas in 1976. Teaching at the University of South Dakota for twenty-two years, he left there in 1996 as Distinguished Regents Professor of Anthropology.

While in South Dakota, he developed a major CRM program and the University of South Dakota Archaeology Laboratory, where he is still a research associate. He was named the University of South Dakota Student Association Teacher of the Year in 1980, given the Burlington Northern Foundation Faculty Achievement Award for Outstanding Teaching in 1986, and granted the Burlington Northern Faculty Achievement Award for Research in 1990. He was selected by Sigma Xi, the Scientific Research Society, as a national lecturer from 1991 to 1993, and he served as executive secretary of the World Archaeological Congress from 1990 to 1994. He has published more than three hundred articles, CRM reports, and reviews and is the author, editor, or coeditor of fifteen books, including *Native North America* (with Brian Molyneaux, University of Oklahoma Press, 2000) and *Indians and Anthropologists: Vine Deloria, Jr., and the Critique of Anthropology* (with Tom Biolsi, University of Arizona Press, 1997). He has served as the editor of *Plains Anthropologist* and the *World Archaeological Bulletin* and as the associate editor of *American Antiquity*. He has done archaeology in the Great Plains of the United States and in Mexico, England, Venezuela, and Australia. He has also worked closely with a wide range of American Indian nations and groups.

William Green is the director of the Logan Museum of Anthropology and an adjunct professor of anthropology at Beloit College, Beloit, Wisconsin. He has been active in archaeology since 1970. Having grown up on the south side of Chicago, he attributes his interest in archaeology and anthropology to the allure of the exotic (i.e., rural) and a driving urge to learn the unwritten past, abetted by the opportunities available at the city's museums and universities. His first fieldwork was on the Mississippi River bluffs in western Illinois. Although he also worked in Israel and England, he returned to Illinois for several years of survey and excavation. His interests in settlement patterns, ceramics, and archaeobotany developed there. He received his master's degree from the University of Wisconsin at Madison and then served as Wisconsin SHPO staff archaeologist for eight years. After obtaining his Ph.D. from the University of Wisconsin at Madison in 1987, he served as state archaeologist of Iowa from 1988 to 2001, directing statewide research and service programs including burial site protection, geographic information, publications, contract services, public outreach, and curation. His main research interests focus on the development and spread of native agriculture. He has served as editor of the *Midcontinental Journal of Archaeology* and *The Wisconsin Archeologist*; has published articles in *American Antiquity, Journal of Archaeological Research*, and other journals; and has received grants and contracts from the National Science Foundation, National Park Service, Iowa Humanities Board, and many other agencies and organizations.